Is Your Picture Worth A Thousand Words?

How to Present Information for Easy Viewing and Understanding

Bill Dommett

Published in Australia in 2018 by Bill Dommett

Email: billdommett@bigpond.com

© Bill Dommett 2018

The moral right of the author has been asserted.

ISBN 9780648271307 (paperback)

Disclaimer

The author has made every effort to ensure the accuracy of the information within this book was correct at the time of publication. The author does not assume and hereby disclaims any liability to any party for any loss, damage or disruption caused by errors or omissions, whether such errors or omissions result from accident, negligence, or any other cause.

To Val, the love of my life.

CONTENTS

List of Tables

List of Figures

Preface

Throughout my career in food laboratories, I was heavily involved in preparing, writing, statistical analysis, editing and presentation of scientific research papers, extension articles, lectures, seminars, reports and work manuals, plus significant experience in quality control, quality assurance and systems analysis and design with and without computers.

Especially during editing of papers, I became aware of many situations where improvements could be made in the tables, graphs and diagrams included, to make them more effective in communicating the key aspects of the work to the readers or audience.

I was also aware of the need to match the text in the methods used, the discussion of results and the conclusions to be drawn with the text and diagrams in the presentation of the results in a logical, easy to follow style.

So I believe there is value in a small book outlining the principles to be aware of when preparing what I have grouped together as *'Pictures'* in the title of this book.

Introduction

Focus and aims of the book

There's an old saying *'A picture is worth a thousand words'*. This may not be literally accurate, but pictures are certainly very valuable aids to communication in all spheres of life. However, to be effective communication aids, pictures need to be made easy to read and interpret. They need to clearly match the accompanying text and focus on the most important aspects of the information contained in both. They also need to be free of any ambiguity and should not confuse or mislead the reader.

If they do not meet these criteria, they may not be effective communication vehicles and in fact may spoil correct interpretation. In other words, they may not be *'worth many words'* at all.

The book is not necessarily only for use by scientific, engineering and technical workers, but should be useful for workers in almost any field. Some of the worst examples of tables, pie charts and other diagrams I have seen were prepared by commercial businesses in their annual reports and marketing material.

In fact, the principles here should be of interest to students preparing assignments and members of clubs and other societies preparing their newsletters and advertising brochures.

Acknowledgement of Microsoft templates used

It should be acknowledged early in this book that to construct many of the diagrams (Figures) in the book, especially from Chapter 2 onward, extensive use has been made of many of the Chart templates publicly available in Microsoft® Word 2013, Microsoft® Word 2016, Microsoft® Excel 2013 and Microsoft® Excel 2016 programs within the Microsoft® Office 2013 and Microsoft® Office 2016 software packages.

In some of the Figures, the tabular data supplied by Microsoft® have been used, but in many others, considerable modifications have been made to the supplied data or different data has been input directly by the author to produce examples of 'desirable' and 'not so desirable' formats.

Definition of some terms used

Throughout this book I will be talking constantly about 'actual data values' or 'data values' or 'values' in a series of tables, graphs or charts. By these shorthand terms, I am really talking about the much longer term 'the appropriate mean value of the actual individual data values, after appropriate statistical analysis of the results'. So please understand that the shorter terms always refer to the much longer term.

Other terms I will be talking about almost as often, especially in the 100% graphs or charts are 'data percentage values' or 'percentage values' or 'percentages'. These terms are also shorthand for the longer real term 'the percentage value of the appropriate mean value of the actual individual data values, after appropriate statistical analysis of the results, and expressed as a fraction of the appropriate total of mean results'. So please understand the real meaning of the term when you read it.

Both sets of shorthand terms are, of course, used to save pages in the book. But also they are employed to focus attention on the appearance, readability and interpretation of the information contained in tables, graphs and charts, which are the main points of the book.

Style and contents

I have tried to write this in a straightforward manner in simple, everyday language, which should be easy to follow and I have included many examples of *'pictures'* (11 tables and 59 charts, graphs and diagrams) to illustrate the points outlined in the text.

Chapter 1 discusses allocation of treatment labels and presentation of information in tables. Many of the principles in Chapter 1 can be applied to non-tabular presentation of information as well, and should be kept in mind as you read each of the following chapters, none of which discusses tables.

Chapter 2 discusses pie charts, with good and bad examples. Chapter 3 focusses on good and bad relative or percentage information presented in column, bar, line or area charts. Chapter 4 is similar to Chapter 3 but focusses on actual data values in the same four chart types.

Chapter 5 applies to scatter or X–Y and X–Y–Z plots in two or three dimensions, while Chapter 6 is about plotting in multiple dimensions (i.e. more than three dimensions) with radar or spider web plots.

Chapter 7 contains special data plotting, such as stock market information, estimation of information spread without statistical analysis and cash (or other asset) flow financial information.

Chapter 8 provides a summary of the book contents as a long bulleted list of points to remember and a few final comments.

So let's begin …

Chapter 1

Choosing the Treatment Labels and Using Tables

Multiple treatments require unique labels before experimentation or observation

If there are a number of treatments involved, each treatment must be assigned a unique label, which can be the full name or short name of the treatment or some combination of code letters or code numbers.

SITUATION 1: Multiple treatments which are independent or unrelated

Treatment labels are to be assigned randomly if the treatments are independent from each other, e.g. if three treatments to be tested for effectiveness of anti-microbial action on equipment surfaces are:

- steam at a set temperature and time,
- disinfectant at a set concentration, temperature and time,
- antibiotic at a set concentration, temperature and time,

then any treatment can be randomly assigned as e.g. (A), any other can be assigned as e.g. (B) and the remaining one can be assigned as e.g. (C).

As will become obvious later in this chapter, these labels can be changed afterwards to optimise presentation of the results and discussion.

SITUATION 2: Multiple treatments which are related or are not independent

If treatments involve increasing or decreasing amounts or levels of the same variable, e.g. various levels of a disinfectant or its application time or its application temperature for anti-microbial action on equipment surfaces, then it would be common practice for labels to be assigned in order of increasing (or decreasing) value of the disinfectant. So if the treatments are:

- 5% chlorine solution at a set temperature and time,
- 10% chlorine solution at same set temperature and time,
- 15% chlorine solution at same set temperature and time,

then commonly the first application would be labelled e.g. (A), the second would be labelled (B) and the last would be labelled as (C).

Again, the labels can be changed later to help results presentation, as explained later in this chapter.

SITUATION 3: Multiple treatments with treatments in groups where treatments within each group are related but treatments are not related between groups

If in the last section, six treatments are to be investigated by adding three applications of iodine at different levels to the three chlorine treatments, such as:

- 5% iodine solution at the same set temperature and time,

- 10% iodine solution at same set temperature and time, and
- 15% iodine solution at same set temperature and time,

then commonly the three additional treatments would be assigned (D), (E) and (F) respectively so (A) and (D) were comparable by each having 5% concentrations, (B) and (E) were similarly comparable with 10% concentrations and (C) and (F) were also comparable.

Again, labels can be changed later for optimal presentation.

Randomising the experimentation or observations

Because a given equipment surface could only be used for one treatment (otherwise the next treatment to be applied to it would encounter fewer microbes to deal with than the number encountered by the preceding treatment), and because of the natural variability of biological data (meaning that the microflora on equipment would vary between items of equipment) replication or repeating of the experiments many times would be required to ascertain the mean or average effect of each treatment and the errors associated with each average or mean value, so treatments could be compared in a practical and statistically valid manner.

There are many different statistical methods and designs to optimise experimentation or observation, none of which will be discussed here, as this aspect is deserving of a full text volume devoted to statistical mathematics. However, in most statistical methods, an assumption of random allocation of treatments is paramount.

So, in any of the situations described above, the treatments would be randomly assigned to reduce effects from other influencing factors such as testing order and time of day (both of which may influence tiredness or boredom of the persons involved and batch or freshness of culture media), day of week (which may influence weather conditions and attitudes of persons involved), experimenter or observer differences (if there is more than one person or more than one laboratory).

In the case of medical experimentation, involving patients and medical examiners, often a 'double blind' experiment (i.e. where the randomising is done by a third party so both the patients and the medical staff are unaware of how the treatment was allocated) is required to eliminate subjective bias.

Analysis and summary of results

Once the experiments or observations are complete, the results are analysed (most commonly with statistical analysis, especially if biological data is involved), summaries and conclusions are made, and the information and appropriate discussion of it are prepared for public or private dissemination. It is here that decisions can be made to optimise the presentation of the information or not.

The temptation now is to present the information in the alphabetical, numerical or alphanumerical order of the treatments involved, i.e. with no changing of labels and with the labels in ascending order. Then the results can be discussed, perhaps in the same order above or perhaps in another order.

In this chapter I will demonstrate an optimising process only with tables of data, but will introduce other presentation vehicles in later chapters.

Optimal information presentations

My strong suggestion at this point is to rearrange the information and discussion to highlight the most important result and present it first, followed by the next most important result, followed by the next most important result, etc until the least important result at the end. This will be demonstrated with the following series of result tables and associated extracts from the methods used in the work and extracts from the discussions/interpretations of results.

[Please note that the data presented are not real data and have been designed only to demonstrate the principles involved.]

EXAMPLE I (With all treatments independent of each other)

Testing 10 chemical treatments for relative anti-microbial effectiveness

Methods extract 1

Treatment A – 5% chlorine* solution at 25°C for 5 minutes;

Treatment B – 5% iodine solution at 25°C for 5 minutes;

Treatment C – 5% hydrogen peroxide at 25°C for 5 minutes;

Treatment D – 5% citric acid at 25°C for 5 minutes;

Treatment E – 5% ethyl alcohol at 25°C for 5 minutes;

Treatment F – 5% methyl alcohol at 25°C for 5 minutes;

Treatment G – 5% benzene at 25°C for 5 minutes;

Treatment H – 5% aspirin at 25°C for 5 minutes;

Treatment I – 5% paracetamol at 25°C for 5 minutes;

Treatment J – 5% ibuprofen at 25°C for 5 minutes.

[*Please note these chemicals are named only to illustrate the principles involved with Example I data and no relative or actual anti-microbial effectiveness is assumed or implied.]

Results table 1

Suppose we have the results of a fictional experiment with the 10 treatments labelled A to J above, which as a first attempt, are displayed in Table 1 below and Discussion extract 1 below:

Table 1: Effectiveness of reducing microbial populations – Percentage of population remaining

Replicate	Treatments									
	A	B	C	D	E	F	G	H	I	J
1	10.0	12.0	15.0	17.0	4.0	5.0	30.0	14.0	42.0	21.0
2	20.0	14.0	16.0	19.0	8.0	10.0	15.0	13.0	34.0	32.0
3	30.0	16.0	14.0	10.0	16.0	11.0	21.0	17.0	43.0	16.0
Sum	60.0	42.0	45.0	46.0	28.0	26.0	66.0	44.0	119.0	69.0
Mean	20.0	14.0	15.0	15.3	9.3	8.7	22.0	14.7	39.7	23.0

Discussion extract 1

The mean percentages of the population remaining after treatments varied widely from 8.7 to 23.0, with individual replicate results varying from 4 to 43. The best result was obtained with Treatment F, with an average percentage of 8.7 and all three replicates below 12. The next best result was that of Treatment E with an average of 9.3 and all three replicates below 17. The next results occurred with a group of treatments including B, H, C and D, which had averages in the range 14.0–15.3. Etc, etc … The worst treatment was I with an average of 39.7.

[COMMENT: The reader would be forced to search up and down and back and forth in the table to follow the discussion which deals with treatments in the order best to worst, while the table has treatments in a different order.]

Suppose we now rearrange the information in Table 1 to match the order of discussion in Discussion extract 1. The rearranged Table 1 is the new Table 2 below and the methods extract and discussion extract remain almost unchanged in Methods extract 2 and Discussion extract 2 below:

Methods extract 2

Treatment A – 5% chlorine solution at 25°C for 5 minutes;
Treatment B – 5% iodine solution at 25°C for 5 minutes;
Treatment C – 5% hydrogen peroxide at 25°C for 5 minutes;
Treatment D – 5% citric acid at 25°C for 5 minutes;
Treatment E – 5% ethyl alcohol at 25°C for 5 minutes;
Treatment F – 5% methyl alcohol at 25°C for 5 minutes;
Treatment G – 5% benzene at 25°C for 5 minutes;
Treatment H – 5% aspirin at 25°C for 5 minutes;
Treatment I – 5% paracetamol at 25°C for 5 minutes;
Treatment J – 5% ibuprofen at 25°C for 5 minutes.

Results table 2

Table 2: Effectiveness of reducing microbial populations – Percentage of population remaining

Replicate	F	E	B	H	C	D	A	G	J	I
					Treatments					
1	5.0	4.0	12.0	14.0	15.0	17.0	10.0	30.0	21.0	42.0
2	10.0	8.0	14.0	13.0	16.0	19.0	20.0	15.0	32.0	34.0
3	11.0	16.0	16.0	17.0	14.0	10.0	30.0	21.0	16.0	43.0
Sum	26.0	28.0	42.0	44.0	45.0	46.0	60.0	66.0	69.0	119.0
Mean	8.7	9.3	14.0	14.7	15.0	15.3	20.0	22.0	23.0	39.7

Discussion extract 2

The mean percentages of the population remaining after treatments varied widely from 8.7 to 23.0, with individual replicate results varying from 4 to 43. The best result was obtained with Treatment F, with an average percentage of 8.7 and all three replicates below 12. The next best result was that of Treatment E with an average of 9.3 and all three replicates below 17. The next results occurred with a group of treatments including B, H, C and D, which had averages in the range 14.0–15.3. Etc, etc … The worst treatment was I with an average of 39.7.

[COMMENT: The reader would be helped considerably with this version because now the table data and the discussion match, so he is no longer forced to search up and down and back and forth in the table to follow the discussion.]

But this situation can be further improved.

After the results are known and the order of effectiveness is known, the treatment assignments in Methods can be rearranged so the treatment code order directly relates to the order in the table and the labels in etc discussion can be altered accordingly. This is illustrated below, in Methods extract 3, Table 3 and Discussion extract 3:

Methods extract 3

Treatment A – 5% methyl alcohol at 25°C for 5 minutes;
Treatment B – 5% ethyl alcohol at 25°C for 5 minutes;
Treatment C – 5% iodine solution at 25°C for 5 minutes;
Treatment D – 5% aspirin at 25°C for 5 minutes;
Treatment E – 5% hydrogen peroxide at 25°C for 5 minutes;
Treatment F – 5% citric acid at 25°C for 5 minutes;
Treatment G – 5% chlorine solution at 25°C for 5 minutes;
Treatment H – 5% benzene at 25°C for 5 minutes;

Treatment I – 5% ibuprofen at 25°C for 5 minutes.
Treatment J – 5% paracetamol at 25°C for 5 minutes.

Results table 3

Table 3: Effectiveness of reducing microbial populations – Percentage of population remaining

Replicate	Treatments									
	A	B	C	D	E	F	G	H	I	J
1	5.0	4.0	12.0	14.0	15.0	17.0	10.0	30.0	21.0	42.0
2	10.0	8.0	14.0	13.0	16.0	19.0	20.0	15.0	32.0	34.0
3	11.0	16.0	16.0	17.0	14.0	10.0	30.0	21.0	16.0	43.0
Sum	26.0	28.0	42.0	44.0	45.0	46.0	60.0	66.0	69.0	119.0
Mean	8.7	9.3	14.0	14.7	15.0	15.3	20.0	22.0	23.0	39.7

Discussion extract 3

The mean percentages of the population remaining after treatments varied widely from 8.7 to 23.0, with individual replicate results varying from 4 to 43. The best result was obtained with Treatment A, with an average percentage of 8.7 and all three replicates below 12. The next best result was that of Treatment B with an average of 9.3 and all three replicates below 17. The next results occurred with a group of treatments including C, D, E and F, which had averages in the range 14.0–15.3. Etc, etc … The worst treatment was J with an average of 39.7.

[COMMENT: Now the Methods, Results table and Discussion are in agreement and the reader is presented with the information in the most practical manner with the best performing treatment in column 1, the intermediate performing treatments in descending order in the middle columns and the least effective treatment in the last column. This arrangement is very practical and easy for the reader to understand and draw appropriate conclusions.]

But the situation can be improved still further.

The table could be even more informative if (a) instead of using code letters in the table headings, the key descriptive word in each treatment was used, or (b) if this is not feasible (because of many narrow columns or many long descriptive words), then a legend explaining the code letters can be placed below the body of the table.

The tables in this section have many narrow columns, so (a) can only be applied by placing treatment names vertically instead of horizontally in the table headings, as in Table 4A below, with matching words in Discussion extract 4 and with no changes to Methods extract 3, now shown as Methods extract 4:

Methods extract 4

Treatment A – 5% methyl alcohol at 25°C for 5 minutes;

Treatment B – 5% ethyl alcohol at 25°C for 5 minutes;

Treatment C – 5% iodine solution at 25°C for 5 minutes;

Treatment D – 5% aspirin at 25°C for 5 minutes;

Treatment E – 5% hydrogen peroxide at 25°C for 5 minutes;

Treatment F – 5% citric acid at 25°C for 5 minutes;

Treatment G – 5% chlorine solution at 25°C for 5 minutes;

Treatment H – 5% benzene at 25°C for 5 minutes;

Treatment I – 5% ibuprofen at 25°C for 5 minutes;

Treatment J – 5% paracetamol at 25°C for 5 minutes.

Results table 4A

Table 4A: Effectiveness of reducing microbial populations – Percentage of population remaining										
Replicate	Treatments*									
	Methyl alcohol	Ethyl alcohol	Iodine	Aspirin	Hydrogen peroxide	Citric acid	Chlorine	Benzene	Ibuprofen	Paracetamol
1	5.0	4.0	12.0	14.0	15.0	17.0	10.0	30.0	21.0	42.0
2	10.0	8.0	14.0	13.0	16.0	19.0	20.0	15.0	32.0	34.0
3	11.0	16.0	16.0	17.0	14.0	10.0	30.0	21.0	16.0	43.0
Sum	26.0	28.0	42.0	44.0	45.0	46.0	60.0	66.0	69.0	119.0
Mean	8.7	9.3	14.0	14.7	15.0	15.3	20.0	22.0	23.0	39.7

*All treatments were applied as 5% solutions at 25°C for 5 minutes

Discussion extract 4

The mean percentage of the population remaining after treatment varied widely from 8.7 to 23.0, with individual replicate results varying from 4 to 43. The best result was obtained with methyl alcohol, with an average percentage of 8.7 and all three replicates below 12. The next best result was that of ethyl alcohol with an average of 9.3 and all three replicates below 17. The next results occurred with a group of chemicals including iodine, aspirin, hydrogen peroxide and citric acid, which had averages in the range 14.0–15.3. Etc, etc ... The worst chemical was paracetamol with an average of 39.7.

Alternatively, using (b), a legend could be added below the table, as illustrated below in Table 4B, which would also be followed by Discussion extract 4:

Results table 4B

Table 4B: Effectiveness of reducing microbial populations – Percentage of population remaining

Replicate	Treatments*									
	A	B	C	D	E	F	G	H	I	J
1	5.0	4.0	12.0	14.0	15.0	17.0	10.0	30.0	21.0	42.0
2	10.0	8.0	14.0	13.0	16.0	19.0	20.0	15.0	32.0	34.0
3	11.0	16.0	16.0	17.0	14.0	10.0	30.0	21.0	16.0	43.0
Sum	26.0	28.0	42.0	44.0	45.0	46.0	60.0	66.0	69.0	119.0
Mean	8.7	9.3	14.0	14.7	15.0	15.3	20.0	22.0	23.0	39.7

*All treatments were applied as 5% solutions at 25°C for 5 minutes

A Methyl alcohol	B Ethyl alcohol	C Iodine	D Aspirin	E Hydrogen peroxide
F Citric acid	G Chlorine	H Benzene	I Ibuprofen	J Paracetamol

[COMMENT: Tables 4A and 4B are far more effective ways of presenting the information in tabular form, compared with the earlier Tables 1, 2 and 3. They make the saying 'A picture is worth a thousand words' almost true.]

However there is one other aspect of this data to consider.

Some workers and readers would consider measuring and displaying anti-microbial effectiveness by the percentage of the population which survives is a 'negative' approach, as the treatment effects are actually on the microbes that are destroyed. They would argue that effectiveness should be measured by the percentage of the population which has been removed or destroyed, so they would regard presenting the data as the percentage destroyed as a 'positive approach' to the work.

Therefore Tables 4A and 4B could be re-worked and presented as Tables 5A and 5B respectively, with the appropriately modified Discussion extract 5 for either table, all as below, with the Methods extract 4 remaining unchanged, but now called Methods extract 5:

Methods extract 5

Treatment A – 5% methyl alcohol at 25°C for 5 minutes;
Treatment B – 5% ethyl alcohol at 25°C for 5 minutes;
Treatment C – 5% iodine solution at 25°C for 5 minutes;
Treatment D – 5% aspirin at 25°C for 5 minutes;
Treatment E – 5% hydrogen peroxide at 25°C for 5 minutes;
Treatment F – 5% citric acid at 25°C for 5 minutes;
Treatment G – 5% chlorine* solution at 25°C for 5 minutes;
Treatment H – 5% benzene at 25°C for 5 minutes;
Treatment I – 5% ibuprofen at 25°C for 5 minutes.
Treatment J – 5% paracetamol at 25°C for 5 minutes;

Results table 5A

Table 5A: Effectiveness of reducing microbial populations – Percentage of population destroyed

Replicate		Methyl alcohol	Ethyl alcohol	Iodine	Aspirin	Hydrogen peroxide	Citric acid	Chlorine	Benzene	Ibuprofen	Paracetamol
						Treatments*					
	1	95.0	96.0	88.0	86.0	85.0	83.0	90.0	70.0	79.0	58.0
	2	90.0	92.0	86.0	87.0	84.0	81.0	80.0	85.0	68.0	66.0
	3	89.0	84.0	84.0	83.0	86.0	90.0	70.0	79.0	84.0	57.0
Sum		274.0	272.0	258.0	256.0	255.0	254.0	240.0	234.0	231.0	181.0
Mean		91.3	90.7	86.0	85.3	85.0	84.7	80.0	78.0	77.0	60.3

All treatments were applied as 5% solutions at 25°C for 5 minutes

21

Discussion extract 5

The mean percentage of the population destroyed by treatments varied widely from 60.3 to 91.3, with individual replicate results varying from 57 to 96. The best result was obtained with methyl alcohol, with an average percentage of 91.3 and all three replicates above 88. The next best result was that of ethyl alcohol with an average of 90.7 and all three replicates above 83. The next results occurred with a group of chemicals including iodine, aspirin, hydrogen peroxide and citric acid, which had averages in the range 84.7–86.0. Etc, etc … The worst chemical was paracetamol with an average of 60.3.

Results table 5B

Table 5B: *Effectiveness of reducing microbial populations – Percentage of population destroyed*

Replicate	Treatments*									
	A	**B**	**C**	**D**	**E**	**F**	**G**	**H**	**I**	**J**
1	95.0	96.0	88.0	86.0	85.0	83.0	90.0	70.0	79.0	58.0
2	90.0	92.0	86.0	87.0	84.0	81.0	80.0	85.0	68.0	66.0
3	89.0	84.0	84.0	83.0	86.0	90.0	70.0	79.0	84.0	57.0
Sum	274.0	272.0	258.0	256.0	255.0	254.0	240.0	234.0	231.0	181.0
Mean	91.3	90.7	86.0	85.3	85.0	84.7	80.0	78.0	77.0	60.3

*All treatments were applied as 5% solutions at 25°C for 5 minutes

A Methyl alcohol   B Ethyl alcohol   C Iodine   D Aspirin   E Hydrogen peroxide
F Citric acid   G Chlorine   H Benzene I Ibuprofen   J Paracetamol

These final Example I items – Methods extract 5, Table 5A or Table 5B and Discussion extract 5 – could therefore be adopted as the best way to present the Example I data to readers, who will be able to grasp the key principles easily and make accurate interpretations and assessments of the work.

[COMMENT: Although re-labelling the treatments in the Methods may look like cheating, it is not, because, even without hindsight, the labelling could have been that way originally as it would have been done arbitrarily. The main point to be made here is that information needs to be presented to the reader in the most effective way to aid understanding.]

EXAMPLE II (With six treatments in two groups of three related treatments to a group)

Testing 6 chemical treatments for relative anti-microbial effectiveness

Following similar logic to that used for Example I, the initial draft methods extract, initial draft table and initial draft discussion extract could be as illustrated in Methods extract 6, Table 6 and Discussion extract 6 below:

Methods extract 6

Treatment A – 5% chlorine solution at 25°C for 5 minutes;
Treatment B – 10% chlorine at 25°C for 5 minutes;
Treatment C – 15% chlorine at 25°C for 5 minutes;
Treatment D – 5% iodine at 25°C for 5 minutes;
Treatment E – 10% iodine at 25°C for 5 minutes;
Treatment F – 15% iodine at 25°C for 5 minutes.

Results table 6

Again, suppose we have the results of a fictional experiment with the six treatments labelled A to F above, which at first attempt, are displayed in Table 6 below:

Table 6: Effectiveness of reducing microbial populations – Percentage of population remaining

Replicate	Treatments					
	A	B	C	D	E	F
1	47.0	27.0	8.0	26.0	13.0	5.0
2	51.0	31.0	10.0	29.0	15.0	6.0
3	53.0	27.0	12.0	31.0	18.0	7.0
Sum	151.0	85.0	30.0	86.0	46.0	18.0
Mean	50.3	28.3	10.0	28.7	15.3	6.0

Discussion extract 6

A first draft extract could be as follows:

The mean percentage of the population remaining after treatment varied widely from 6.0 to 50.3, with individual replicate results varying from 5 to 53. The best mean result was obtained with Treatment F, with an average percentage of 6.0 and all three replicates below 8. The next best mean result was that of Treatment C with an average of 10.0 and all three replicates below 13. The next results occurred with treatments in the order E, B and D, which had averages of 15.3, 28.3 and 28.7 respectively. Etc, etc … The worst treatment was A with an average of 50.3.

[COMMENT: The reader again would be forced to search up and down and back and forth in the table to follow the discussion which deals with treatments in the order best to worst, whereas the table is in a different order.]

A potential problem now is that the overall effectiveness in descending order is: F > C > E > B > D > A, which really means: 15% Iodine > 15% Chlorine > 10% Iodine > 10% Chlorine > 5% Iodine > 5% Chlorine, so should the data be displayed in the table in this overall manner, ie keeping the solution strengths together, with the chemicals spread widely? Or should the chemicals be kept together, with the strengths spread throughout?

There are advantages to either approach, depending on the use to which the information will subsequently be put. However, there would be major differences between the two chemicals in cost, availability, storage life, residual presence on equipment and human and animal safety, all of which will affect their use in practice. Naturally, strengths will impact on some of these factors too.

Basically, I believe there are two objectives to be achieved in this example – the effect of different chemicals and the effects of chemical strengths. Of these two, I believe the differences between chemicals would be the primary objective and differences between chemical strengths to be a secondary objective as a sort of 'fine tuning' for each of the chemical effects.

Therefore I would keep all the data for each chemical together, ie keep Treatments A, B and C together in adjacent columns and keep Treatments D, E and F together in adjacent columns. This is because the effects of strength would be expected to increase as strength increased, and the results will only quantify this expectation, whereas the effects of chemicals cannot be so easily expected and would provide the most important information, which can then be considered along with other factors above such as cost, safety, etc.

I would also address both of these objectives separately in the discussion – with effects of chemicals illustrated first with assistance from extra data such as overall mean chemical results and with effects of strengths illustrated second.

Skipping some of the intermediate steps outlined with Example I, my next draft (with hindsight) would label the iodine treatments first in the Methods extract (as the better performing group of treatments, and therefore the more important set of results). These would be put in decreasing strength order as A, B and C, followed by the chlorine treatments also in decreasing strength order as D, E and F.

My next draft of the table would have treatments in the same matching order, as in the new Methods extract and my next draft of the Discussion extract would be markedly changed to separate the effects of chemicals from the effect of chemical strengths.

These changes are illustrated in Methods extract 7, Table 7 and Discussion extract 7 below:

Methods extract 7

Treatment A – 15% iodine solution at 25°C for 5 minutes;
Treatment B – 10% iodine at 25°C for 5 minutes;
Treatment C – 5% iodine at 25°C for 5 minutes;
Treatment D – 15% chlorine at 25°C for 5 minutes;
Treatment E – 10% chlorine at 25°C for 5 minutes;
Treatment F - 5% chlorine at 25°C for 5 minutes.

Results table 7

Note that this also contains some extra data, e.g. the overall mean values for the average residual population for each of the chemicals.

Table 7: Effectiveness of reducing microbial populations – Percentage of population remaining

Replicate	Treatments (all conducted at 25°C for 5 minutes)						
	15% Iodine	10% Iodine	5% Iodine	15% Chlorine	10% Chlorine	5% Chlorine
1	5.0	13.0	26.0	8.0	27.0	47.0
2	6.0	15.0	29.0	10.0	31.0	51.0
3	7.0	18.0	31.0	12.0	27.0	53.0
Sum	18.0	46.0	86.0	30.0	85.0	151.0
Mean	6.0	15.3	28.7	10.0	28.3	50.3
Overall mean for chemical	16.7			29.5		

Discussion extract 7

The mean percentage of the population remaining after treatment varied widely from 6.0 to 50.3, with individual replicate results varying from 5 to 53. Iodine was the better of the two chemicals tested; its overall mean percentage was 16.7 compared with 29.5 for chlorine.

The best mean iodine result was 6.0 with 15% solution; 15.3 was obtained with 10 % solution and 28.7 was obtained with 5% solution. The best mean result with chlorine was 10.0 with 15% solution; 28.3 was obtained with 10% solution and the worst result was 50.3 with 5% solution.

With both chemicals, residual population decreased as the strengths were increased. ... Etc, etc ...

However, as with Example I, these data may perhaps be regarded as part of a 'negative' approach, so they could be changed to present a more 'positive' approach to the work, by using percentage of population destroyed instead of percentage of population surviving. This would result in no further changes to the Methods extract, but marked changes to the Results table and the Discussion extract.

These changes have been incorporated in Results table 8 and Discussion extract 8 below, with Methods extract 7 repeated but now called Methods extract 8:

Methods extract 8

Treatment A – 15% iodine solution at 25°C for 5 minutes;
Treatment B – 10% iodine at 25°C for 5 minutes;
Treatment C – 5% iodine at 25°C for 5 minutes;
Treatment D – 15% chlorine at 25°C for 5 minutes;
Treatment E – 10% chlorine at 25°C for 5 minutes;
Treatment F – 5% chlorine at 25°C for 5 minutes.

Results table 8

Results table 8

Table 8: Effectiveness of reducing microbial populations – Percentage of population destroyed

Replicate	Treatments (all conducted at 25°C for 5 minutes)					
	15% Iodine	10% Iodine	5% Iodine	15% Chlorine	10% Chlorine	5% Chlorine
1	95.0	87.0	74.0	92.0	73.0	53.0
2	94.0	85.0	71.0	90.0	69.0	49.0
3	93.0	82.0	69.0	88.0	73.0	47.0
Sum	282.0	254.0	214.0	270.0	215.0	149.0
Mean	94.0	84.7	71.3	90.0	71.7	49.7
Overall mean for chemical	83.3			70.5		

Discussion extract 8

The mean percentage of the population destroyed after treatment varied widely from 49.7 to 94.0, with individual replicate results varying from 47 to 95. Iodine was the better of the two chemicals tested; its overall mean percentage was 83.3 compared with 70.5 for chlorine.

The best mean Iodine result was 94.0 with 15% solution; 84.7 was obtained with 10 % solution and 71.3 was obtained with 5% solution. The best mean result with chlorine was 90.0 with 15% solution; 71.7 was obtained with 10% solution and the worst result was 49.7 with 5% solution.

With both chemicals, the proportion of the population destroyed increased as the strengths were increased. ... Etc, etc ...

As with Example I, we could adopt these final items – Methods extract 8, Table 8 and Discussion extract 8 – as elements to be included in the final presentation of the data, as the combination of these will present the easiest version of the data for the reader to see and understand.

After all we don't want to find that the most important result in a table is buried in the 4th row of the 3rd column from the right and therefore is hard to find to relate to a logical arrangement and order of the discussion!

[Again, no cheating has occurred in this process, which has however produced benefits for the reader.]

Chapter 2

Using Pie Charts

Pie chart use

When the items of a data set can be divided into various categories which make up the whole set, pie charts can be used to graphically indicate the proportions of each of the categories in the set. This is often done by calculating the percentages for each category, and representing the original data and/or the percentages as segments of a circular 'pie'.

Pie charts can be two-dimensional circles rather like a clock face (Figure 1) or a hollow ring (Figure 2) or can be three-dimensional columns like a tilted clock face with minimal thickness (Figure 3), as shown below, in which the same data set is displayed in three different ways:

Pie

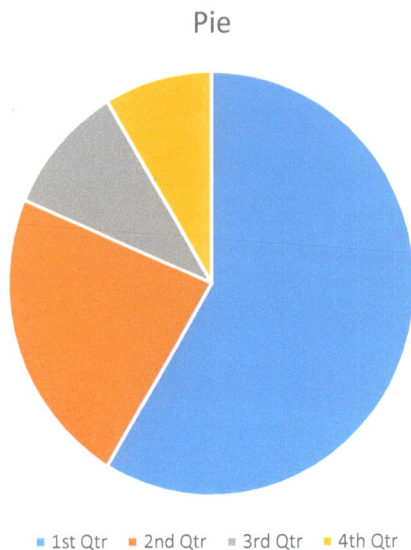

Figure 1: Sales over one financial year

The data which produces each of these charts can be in a table of values in a Word document or an Excel worksheet, here shown as Table 9:

	Sales
1st Quarter	8.2
2nd Quarter	3.2
3rd Quarter	1.4
4th Quarter	1.2

Table 9: Sales by quarters

Donut

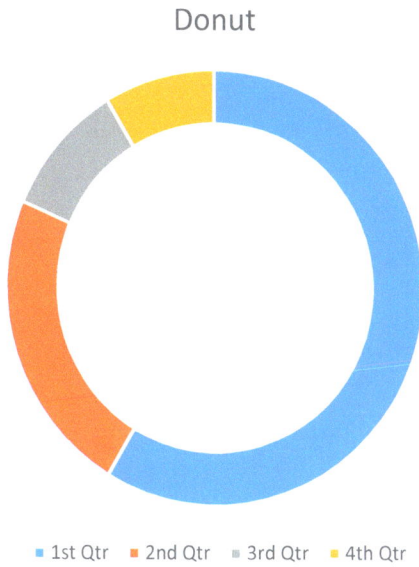

1st Qtr   2nd Qtr   3rd Qtr   4th Qtr

Figure 2: Sales over one financial year

3D Pie

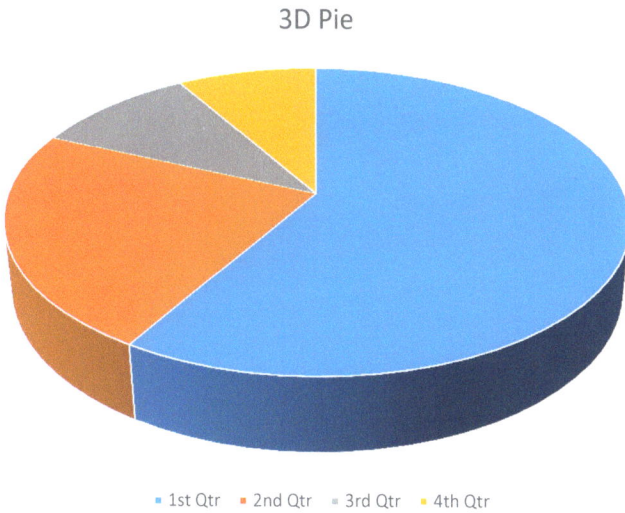

1st Qtr   2nd Qtr   3rd Qtr   4th Qtr

Figure 3: Sales over one financial year

Please note that Figures 1, 2 and 3 are all equivalent because in each, the angle of the first segment is the same fraction of 360° (or of π radians) for all three charts, the angle for the second segment is the same fraction of 360° (or of π radians) for all charts, etc, so the ratios between data values are mirrored in the ratios between segment angles in the chart and the ratios between segment arcs in the charts.

It is helpful to have the values for each of the segments included in the chart, as this not only gives the reader a visual presentation but the actual values as well. This is demonstrated in Figure 4, which is Figure 1 repeated with the values.

Pie

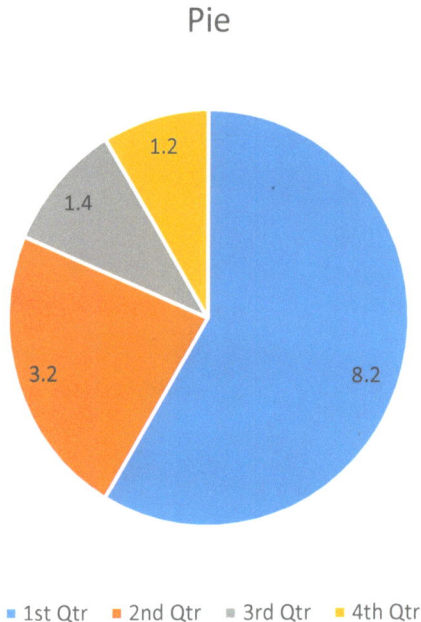

■ 1st Qtr   ■ 2nd Qtr   ■ 3rd Qtr   ■ 4th Qtr

Figure 4: Sales over one financial year

Please also note that for pie charts to be compared within a document or between documents, the starting point or 'origin' should be at the very top of the circle, i.e. in what would be the 12 o'clock position on a real clock face, and the categories should appear in a clockwise direction, i.e. from 12 o'clock to 1 o'clock to 2 o'clock, etc until the last of the categories finishes at the 12 o'clock position again.

Categories should appear in descending order as you read the graph in a clockwise manner, i.e. with the largest starting at the 12 o'clock position and the smallest finishing at the 12 o'clock position.

Inferior pie chart designs

Occasionally, one may encounter pie charts which do not start at the top (12 o'clock position) and these are a little more difficult to read and interpret, so these styles should not be used in the interests of uniformity and ease of interpretation, as illustrated with the same data from Figure 1 in Figure 5, which has been rotated 70° to the right, i.e. clockwise.

More often, one will encounter pie charts in which the categories are not displayed in descending order, but are (a) almost randomly ordered according to data values but perhaps ordered according to some other factor, or (b) perhaps ordered in a pattern of 'a large category-followed by a small category-followed by another large category-followed by another small category' to cause the small values to be displayed separately rather than all together at the end of the circular pattern. Both (a) and (b) are illustrated in Figures 6 and 7 respectively below, both of which contain different data from that in Figures 1–4.

In Figure 6, the largest and second largest categories are Shop 4 and Shop 5 respectively, but you have to look all over the diagram to realise this fact; the smallest and the second smallest categories are Shop 2 and Shop 3 respectively, which you have to look at the diagram closely to appreciate; also Shop 1 and Shop 8 are very close in size but Shop 1 is actually a slightly larger category than Shop 8.

If entries were in descending order, these observations would be easier and quicker to understand, and would match discussion of results if this is also so arranged, starting with the most important and finishing with the least important. See Chapter 1 for a fuller explanation of the desirability to have the Methods, Results and Discussion all matching and, in most cases, all focussed on the most important information first and the least important information last.

Sales

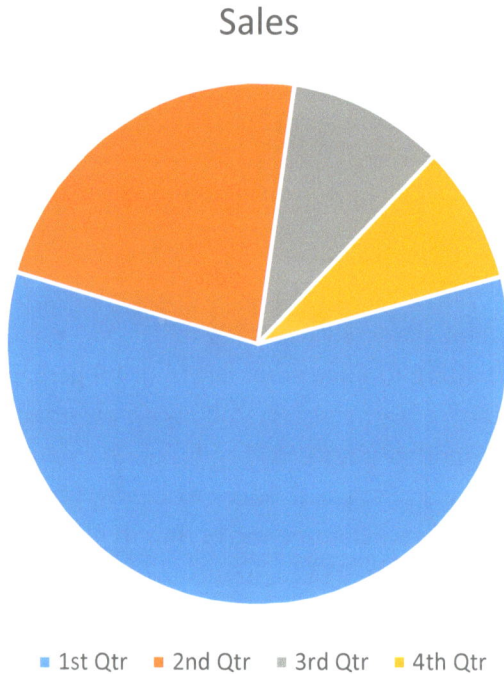

■ 1st Qtr ■ 2nd Qtr ■ 3rd Qtr ■ 4th Qtr

Figure 5: Sales over one financial year with largest segment not starting at the 12 o'clock position

Sales

Figure 6: Sales over one financial year with large and small segments in mixed order

In Figure 7, which has different data again from that in Figure 6, the situation is even worse, as the reader has to search the diagram at length and many times to follow the normal order in which discussion of results would most likely occur. Here the descending order of data values is with Shops in the order: 1, 3, 7, 5, 8, 6, 4 and 2.

Neither of these modified order pie charts are very useful, because they do not display results in a best to worst fashion, which would be the normal way of discussing results, i.e. dealing with the most important or significant result first and then dealing with less important results in descending order. So it is best if results figures are already in descending order to make following the discussion easier and logical for the reader.

Sales

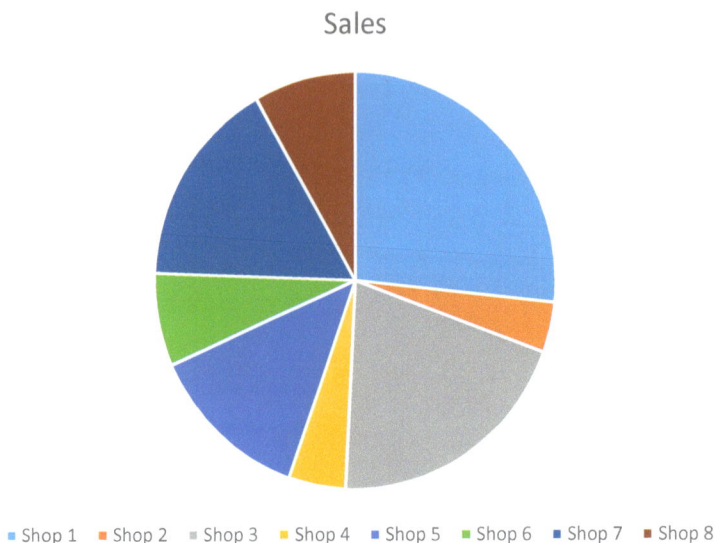

Figure 7: Sales over one financial year with large and small segments deliberately alternating

Key differences from coordinate geometry

In two-dimensional coordinate or analytical geometry, which is a part of modern mathematics, figures like circles, triangles and angles are drawn on a grid with two axes; the X axis is drawn in the horizontal direction, with positive values on the right and negative values on the left, and the Y axis is drawn in the vertical direction, with positive values towards the top and negative values towards the bottom. The axes cross each other at the origin, where values of X and Y are both zero. See Figure 8 below:

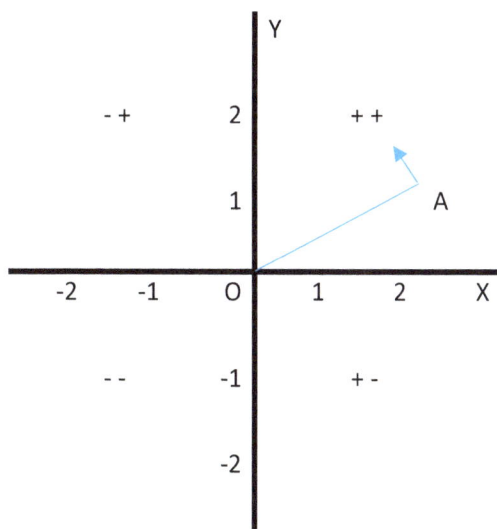

Figure 8: Coordinate geometry example, with rotating line and angle

Angles and segments of a circle start from the horizontal axis X on the right (i.e. the 3 o'clock position in a pie chart, unlike the preferred 12 o'clock normally used) and rotate anti-clockwise (unlike ideal pie charts which rotate clockwise).

The graph is divided by the two axes into four parts called quadrants. The X and Y values are both positive in the first quadrant (top right quadrant) as indicated by the + + symbols; so angles less than or equal to 90° (i.e. acute angles) have positive values for both X and Y.

Going around the origin in an anti-clockwise direction, values of X and Y change to negative as indicated by the - +, - - and + - symbols in each quadrant; therefore angles greater than 90° but not greater than 180° (i.e. obtuse angles) have negative X and positive Y, and so on around the quadrants.

Alternative pie charts

If the smallest values in a pie chart are difficult to read, there are a few ways that their values can be differentiated – both of which use an expansion or magnification of the real values in another small pie chart or bar chart to indicate their *relative* strengths only. These are called Pie of Pie and Bar of Pie charts, as illustrated in Figures 9 and 10 below. However, neither of these diagrams has the primary plot or the secondary plot starting at the ideal 12 o'clock position:

Pie of Pie

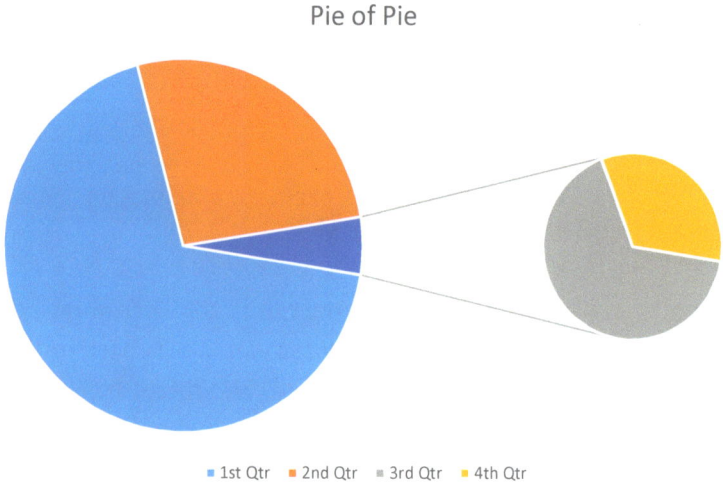

■ 1st Qtr   ■ 2nd Qtr   ■ 3rd Qtr   ■ 4th Qtr

Figure 9: Alternative pie chart with secondary plot to differentiate amongst the smallest values

Alternatives to colour to indicate categories

Colour is the most effective way of identifying categories, especially if sufficient contrast exists between adjacent categories, as evidenced by all the pie charts illustrated above. If there are two or more coloured diagrams in a document (pie charts or any of the charts in the following chapters), it is highly desirable that the same colour scheme is applied to all, as this assists interpretation of the information by readers.

If non-contrasting colours are used for adjacent categories, especially if there are many categories, such as the different shades of blue in Figure 11, ease of reading the information can be compromised.

Bar of Pie

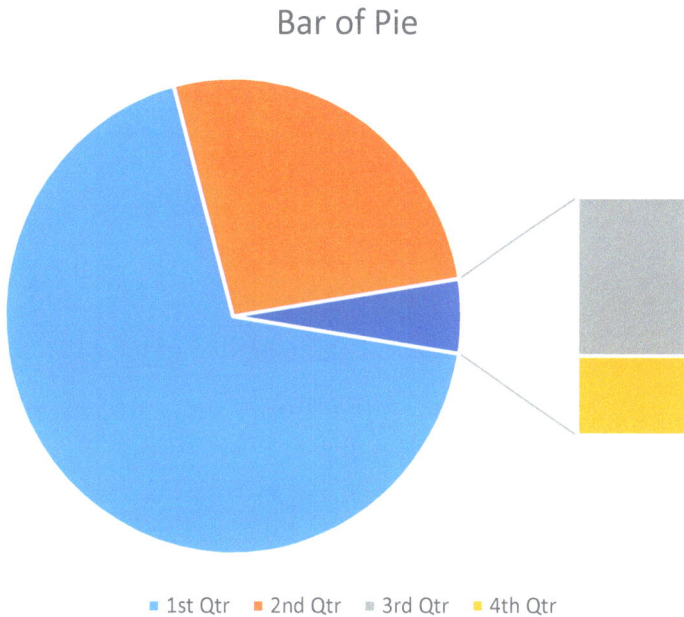

Figure 10: Another alternative pie chart with a secondary plot to differentiate among small values

Sales

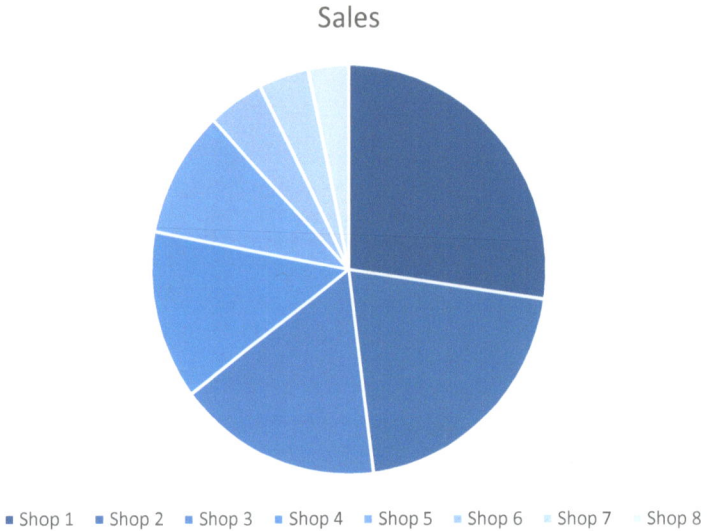

■ Shop 1 ■ Shop 2 ■ Shop 3 ■ Shop 4 ■ Shop 5 ■ Shop 6 ■ Shop 7 Shop 8

Figure 11: Pie chart with non-contrasting colours spoiling easy identification of categories

Another option, if colours other than black, grey and white are not able to be used, is to use different shades of grey as in Figure 12 or black stripes, hatch marks or dot patterns to separate the categories as in Figure 13. However both of these are poor substitutes for contrasting colours.

If the appropriate selection and formatting of pie charts is done and the appropriate arrangement of the data is made, then it should aid readers trying to understand the information in the data. Therefore if you are guided by the principles included in this chapter and in the previous chapter, your pie chart diagrams should confirm the expression that: 'A picture is worth a thousand words.'

Sales

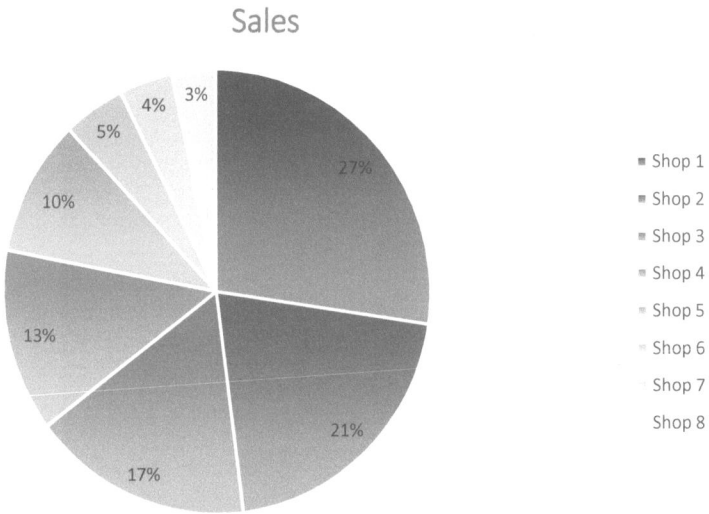

- Shop 1
- Shop 2
- Shop 3
- Shop 4
- Shop 5
- Shop 6
- Shop 7
- Shop 8

Figure 12: Pie chart only in shades of grey for black and white printing

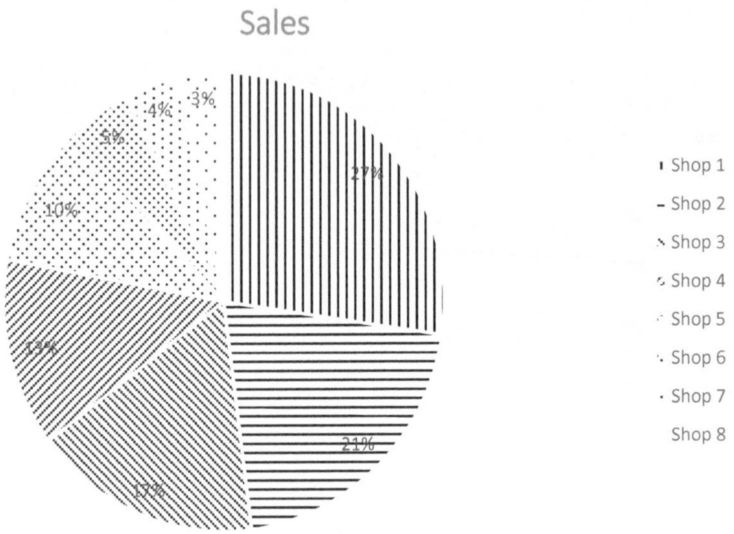
Sales
27%
21%
17%
13%
10%
5%
4%
3%
Shop 1
Shop 2
Shop 3
Shop 4
Shop 5
Shop 6
Shop 7
Shop 8

Figure 13: Pie chart only in grey patterns for black and white printing

Chapter 3

Using Column, Bar, Line and Area Graphs with Relative Data Changes

Columns and bars

The terms 'column graph' and 'column chart' are normally restricted to graphs or charts where the data is presented in a vertical format, rather like columns or pillars often found at the entrance of large buildings as decorations. In a 'bar graph' or 'bar chart', the data can be presented vertically or horizontally, but often this term is restricted to graphs where the data is presented horizontally, rather like a metal bar used to hang curtains from, so as to have specific names for each format.

In this chapter, discussion will focus on the vertical presentation of data, i.e. as columns, which I prefer, but all of the comments could be equally applied to the horizontal format, i.e. as bars, under the stricter definition above. Examples of both types are presented as Figures 14 and 15 below.

Clustered Column

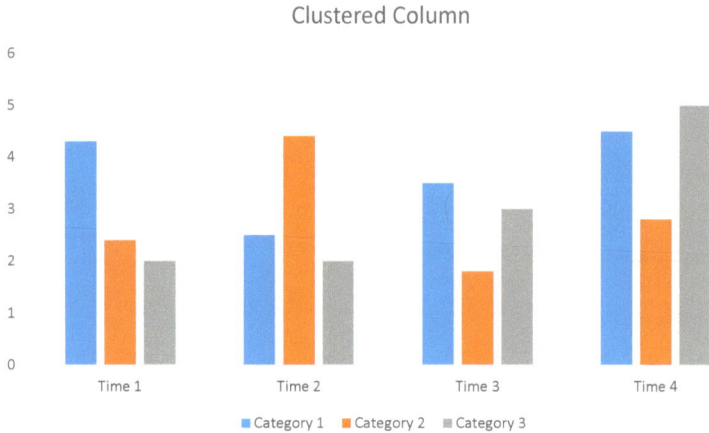

Figure 14: Example of a column chart or graph

Histograms used to display categories

Some column charts can be used as the pie charts discussed in Chapter 2 are used – they can present a population of data in categories, either as values or as percentages. One of the most common of these is the histogram, which typically divides the data for a perhaps otherwise homogeneous population into artificial groups or ranges and then plots the proportions of the data which occur in each of the ranges.

Clustered Bar

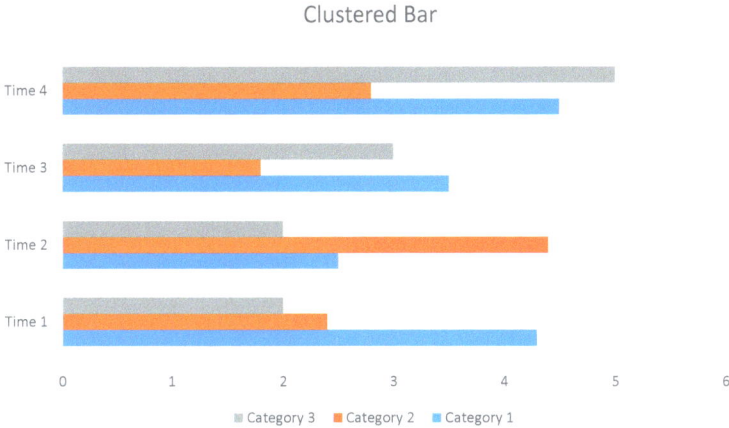

Figure 15: Example of a bar chart or graph, using the same data as in Figure 14

A good example would be to sort the human population of an area (perhaps a city, a district, a state or a nation) into age groups and plot the numbers or percentages occurring in each age group. It is important to be clear about the boundaries of each group so as not to introduce ambiguity or errors in the presentation.

For example, if one age group is defined to be children under 5 years (or <5) and the next age group is defined as children over 5 but under 10 (or >5 to <10), there is no category for children who are 5. So children who are five will have to be allocated to one specific group. This can be done in two ways: (a) age groups could be 0–4 (or <5) and 5–9 (or 5–<10), or (b) 0–5 (or <6) and 6–10 (or >5 – 10). Whichever choice is made with these first two categories, the same choice should be made with each of the other categories, so the population is split as evenly as possible.

Example of errors with histograms

An example of a simple histogram is displayed in Figure 16 below. However, there are errors in the labelling of the horizontal axis, because boundary values could be assigned to either of the categories associated with the boundary, e.g. the first range is called (1, 5) [or 1–5] and the second range is called (5, 9) [or 5–9]. This type of error where 5 belongs in both categories has just been discussed above.

The number in each coloured block is the number of occurrences in that category. Perusal of the data here plotted shows that it consists of integers only and the lowest 17 actual values are: 1, 3, 3, 3, 5, 6, 6, 6, 7, 8, 8, 9, 9, 9, 9, 9 and 10, which means there are five values in the range 1–5, followed by 11 values in the range 6–9, followed by several values in each of the higher ranges.

So the fifth value, 5, has actually been placed with the 1 and the three 3's into the category labelled correctly as (1, 5) [which could be optionally labelled as 1-5]. The next 11 values, i.e. three 6's, one 7, two 8's and five 9's have been placed into the category incorrectly labelled as (5, 9) [meaning 5–9], but which should have been labelled as (6, 9) [or 6–9]. The next value of 10 and 22 other values have been placed in the next category incorrectly labelled as (9, 13), which should be (10, 13) [or 10–13].

Histogram

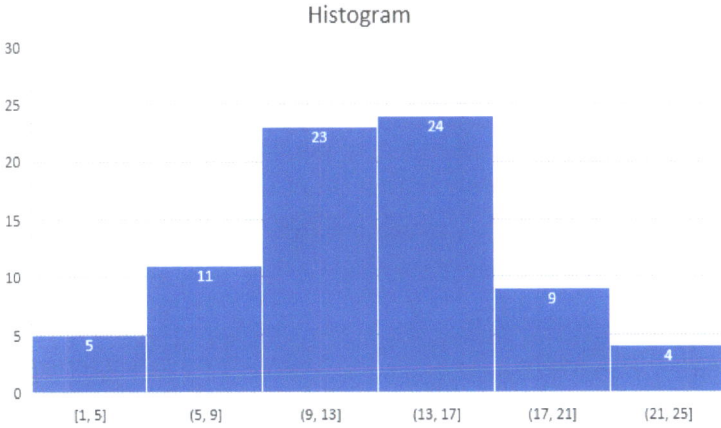

Figure 16: Histogram of young age groups in an area (with errors in the horizontal axis labelling)

The remaining incorrectly labelled axis markings should be, in order, (14, 17), (18, 21) and (22, 25) [or optionally 14–17, 18–21 and 22–25].

If the plotted data values had included fractional values instead of only whole numbers, then the categories, in order, would need to have been labelled as (0, 5), (>5, 9), (>9, 13), (>13, 17), (>17, 21) and (>21, 25) like those in the diagram [or perhaps as 0–5, >5–9, >9–13, >13–17, >17–21 and >21–25].

Of course, if colour production is not available to be used, then black, grey and white would need to be used instead of the blue in Figure 16.

100% stacked column charts

Many times, category values in a data population will change over time or before and after treatments, so different graph or chart types are required to demonstrate these changes.

Figure 17 demonstrates one of these, which is called a 100% Stacked Column chart, as it displays the relative or percentage strengths of each category at each of a series of times.

Note that the numbers in the coloured blocks are the actual values for the categories at each time, e.g. at Time 1, the blue category value was 4.3, which is 49.4% of 8.7, the total of all categories at Time 1.

In keeping with the principles outlined in previous chapters, especially Chapter 1 concerning tables, I would recommend placing the most important category in the first or lowest position of any 100% Stacked chart, as it is the easiest position in which the percentages can be read directly off the grid lines. The legend below the graph will also direct readers to the most important category if it is placed lowest on the percentage scale and first in the legend.

All other categories require reading two grid marks and subtracting to obtain the correct percentages. Also, readers are more likely to examine the graph from either the bottom, i.e. nearest the horizontal axis, or the top, and not from any intermediate graph plot positions. I certainly would not recommend placing the most important category or information in the intermediate plot positions.

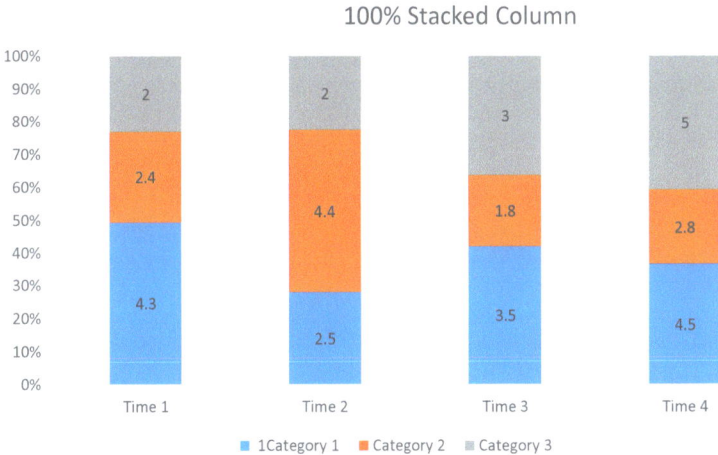

Figure 17: 100% Stacked Column Chart, illustrating category percentage changes over time

(Note that numbers in each coloured block are the actual values, not the percentages, for example the Category 1 value of 4.3 at Time 1 is 49.4% of the total value of 8.7 at Time 1)

Other 100% stacked charts

100% Stacked charts also exist for Bars (which is basically Columns expressed horizontally), and also for Lines and for Areas. A 100% Line chart and a 100% Area chart are both shown below as Figures 18 and 19 respectively, both of which display another aspect of the data in Figures 14 and 15, and in the case of Figure 19, there is a fifth set time added, with a repeat of the same data as at the fourth set time.

Please note that despite Figure 19 being called an Area chart, Figures 17, 18 and 19 are all actually plotting heights as a percentage of the total height at a given time, rather than plotting areas as a percentage of the total area at a given time. Because of the uniform widths of the columns in Figure 17, height ratios in it would be directly related to area ratios anyway, and because there are no widths in Figure 18, areas are not even evident or relevant.

However, close examination of Figures 18 and 19 would suggest there are linear changes occurring in all categories between times, as there are straight lines linking the data plotted at each of the times, and this may not necessarily be true, even for any one of the categories. For this reason, I tend not to use 100% Stacked line graphs or 100% Stacked Area graphs unless it has been established that there are in fact linear changes involved.

Another undesirable aspect of 100% Stacked Area graphs in my opinion is that there are quadrilateral shaped areas for each category between times, which visually tends to emphasise the concept of an 'area' in this chart type, and may cause the reader to mistakenly think in terms of areas between readings instead of the height readings at set points only. We actually know nothing about categories between readings, as shown correctly in Figure 17!

It is worth mentioning that 3-dimensional versions of Figures 17, 18 and 19 are available, but apart from looking attractive, they add nothing extra to be discussed in the situation being examined here than the 2-dimensional versions. A 3-Dimensional 100% Stacked Column version of Figure 17 is displayed as Figure 20 below. Similar 3D versions of 100% Stacked Lines and 100% Stacked Areas exist.

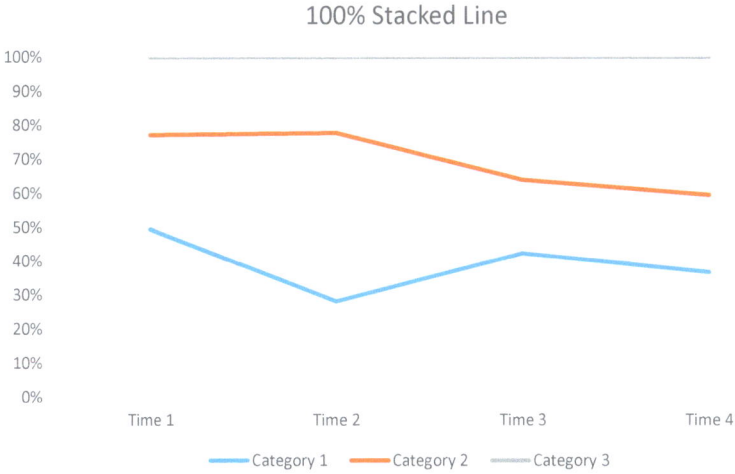

Figure 18: 100% Stacked Line graph showing category percentages at set times

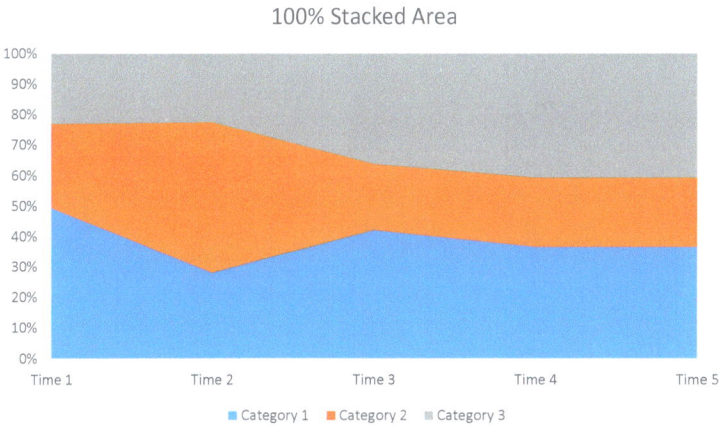

Figure 19: 100% Stacked Area chart showing category percentages at set times

Again, although Figure 20 is 3-dimensional, it is still strictly the height being plotted, not the area or the volume, although in the case of 100% Stacked Columns, because of the uniform horizontal cross-sectional area of the columns or bars, the ratio of heights is equal to the ratios of the areas or the ratios of the volumes. Similar comments could be made about the 3D versions of 100% Stacked Bars, 100% Stacked Lines and 100% Stacked Areas.

So, from all the comments in this chapter, not only can graphs behave as 'pictures worth a thousand words' when properly presented, but they can, if improperly constructed, visually convey 'ghost words or concepts' that are not present and may not actually exist.

3D 100% Stacked Column

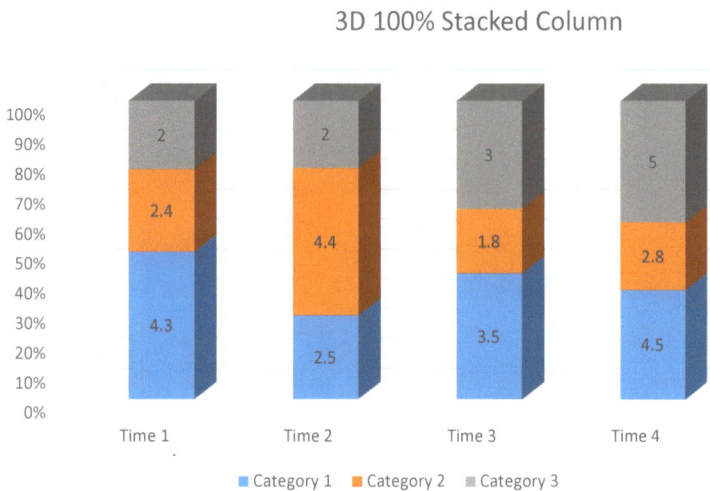

Figure 20: 3-Dimensional version of the 100% Stacked Column chart in Figure 17

Chapter 4

Using Column, Bar, Line and Area Graphs for Actual Data Changes

Whereas in Chapter 3 we dealt with histograms for plotting frequencies and various 100% stacked charts for plotting percentage values of categories, in this chapter we will deal with various charts for plotting actual values. Chart types can include Columns, Bars, Lines and Areas, but as explained in the previous chapter, Bars are the same as Columns except that they are plotted horizontally, so they will not be discussed separately, as any comments for Columns can be applied to Bars as well.

Column charts

The first chart to be discussed is the Clustered Column chart in Figure 21 which displays the actual result values, not the percentages. It is Figure 14 repeated, but with standard error or standard deviation vertical bars on either side of the mean to indicate the spread of results. Error bars will be eliminated from most following graphs to reduce clutter, but please assume they are desirable.

Clustered Column

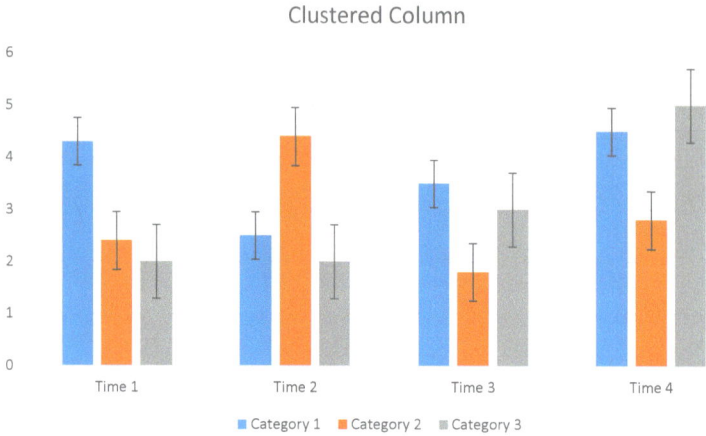

Figure 21: A column chart or graph displaying changes in the category values and standard errors over time

Again, in line with the principles already established in earlier chapters, I would recommend placing the most important category in the first position of the group at each time. In the figure, I have assumed it to be the blue columns, because the blue category is the dominant one at the start and over all times has the highest total (14.8), but over time it is overtaken by the grey category.

Other workers may assume that the grey columns are the most important, because although they are the weakest at the start, they dominate at the end.

So decisions on placing categories in order from most important to least important depends on the appropriate point of view and therefore it is subjective, but it should be better and hopefully more meaningful than any random order of the results.

Discussion of the results also should start with the most important information to be gained from the diagram of results, and should match the structure in the diagram so the reader can follow the arguments in the discussion and see the appropriate results in the diagram easily.

Naturally, if there is only a single category involved, the diagram above loses two of the colours and order is not involved at all.

Three-dimensional versions of Figure 21 are available and two versions appear below as Figures 22 and 23, with the same values contained in Figure 21.

3D Clustered Column

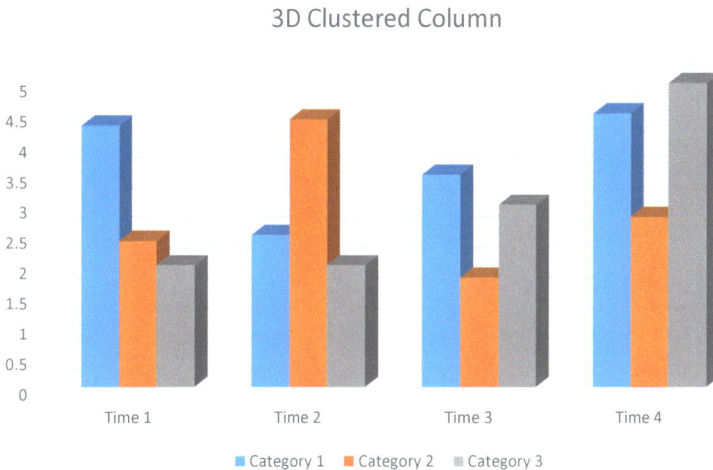

Figure 22: 3-Dimensional Column chart with the same actual result values as in Figure 21

Both charts look attractive but do not add any additional information from that in Figure 21. Figure 22 is styled in the same grouped category format used in Figure 21 where all the data is visible, so that none of the values are hidden or obscured in Figures 21 and 22.

Figure 23 has the categories plotted in a third dimension, making a 2X2 matrix view of categories by times possible. This has the advantage of seeing the changes in any category (i.e. any colour) through time, without the other categories (i.e. other colours) cluttering the in-between spaces, as occurs in the 3D clustered column when viewing Figures 21 or 22 from left to right.

However Figure 23 has the disadvantages of obscuring or hiding some of the smaller columns that are behind taller ones and some difficulty in reading the actual heights of some of the columns, especially three of the red columns in the middle right and three of the blue columns in the front right.

As discussed in Chapter 3, in all the 3D graphs, including more to follow, the values are plotted as heights, not areas or volumes. While the uniform cross-sectional area in a Column or Bar graph makes ratios of heights equal to ratios of areas and to ratios of volumes, areas and volumes are absent in Line graphs, but the presence of areas and volumes in Area graphs can be confusing.

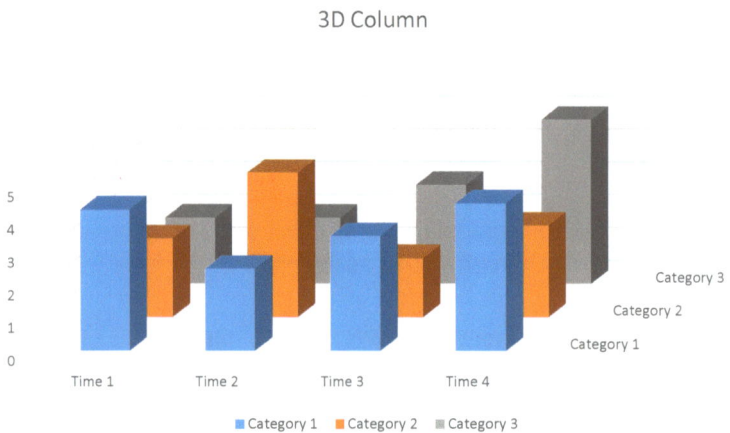

Figure 23: Another 3-Dimensional column chart of the actual results values from Figure 21

Stacked column charts

Figures 24 and 25 show how data values can be stacked one above the others to display not only the plotted height of each value but the total of the stack of values at each time. Figure 24 is 2-dimensional and Figure 25 is the 3D version of the same data. Please note these stacks include the actual values, not the percentage values discussed in Chapter 3. So you will need to decide which format suits your purposes.

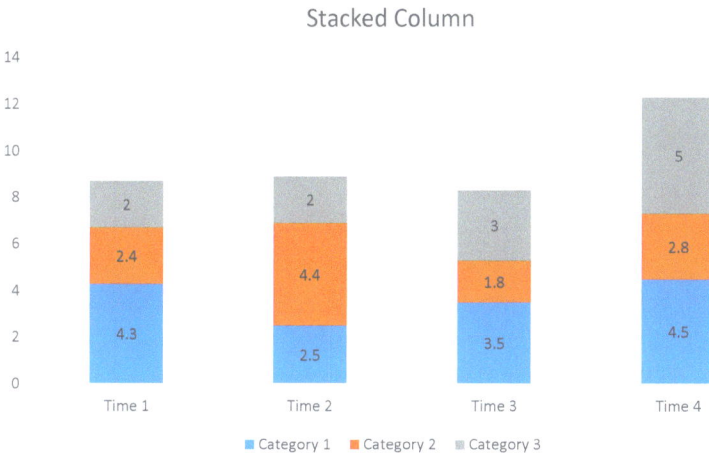

Figure 24: Stacked column chart of the same data in Figure 21, showing individual and total values

3D Stacked Column

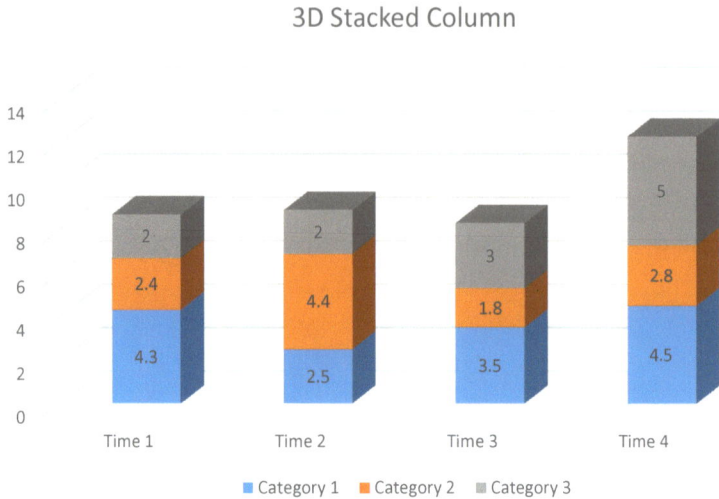

Figure 25: 3D version of Figure 24, still containing the data in Figure 21

The only extra information in these two graphs is that the four totals can now be read off the chart. The 3D version again is attractive but supplies no extra information beyond that in the 2D version. In fact, in the 3D version, it is difficult to judge the total heights of columns against the background grid.

Again, I would recommend the most important information would be best placed as the first category, i.e. the blue sections in these examples, and certainly not in the intermediate sections, i.e. the red sections.

Line charts

Instead of column or bar charts, line charts or area charts can be chosen, and 2D versions and 3D versions of both of these are available. Figure 26 displays a simple Line Chart without markers for the actual data points, while Figure 27 is the same but with markers added, plus standard error bars.

I prefer the version with the markers, indicating there are only four values for each category, but the major fault in both of these is that the lines connecting the point values could suggest that there is a linear change between values, which may or not be true. If true, then the graphs are indicating real linear changes between times; if not true, then Column or Bar charts would be more appropriate models to use, as these do not try to indicate what happens between readings.

A 3D version of line charts is available, but I think it is far too difficult to read and so I would not recommend it. It is displayed in Figure 28.

Stacked line charts

Stacked versions of Line charts exist, and one is shown in Figure 29. It is not difficult for a reader to be confused and think this is a graph of all actual values, each plotted separately from the horizontal axis rather than a stacking of three sets of data, one above each other. In other words, they may read the first red value as over 6, when it is really 2.4 and they may read the first grey value as over 8, when it is really 2.0. The only clue for the reader is the word 'Stacked' appearing twice in the chart.

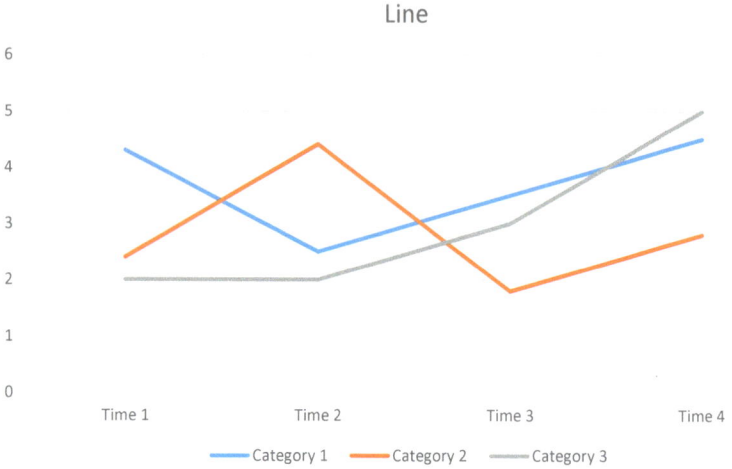

Figure 26: Line Chart of the same data shown as columns in Figure 21

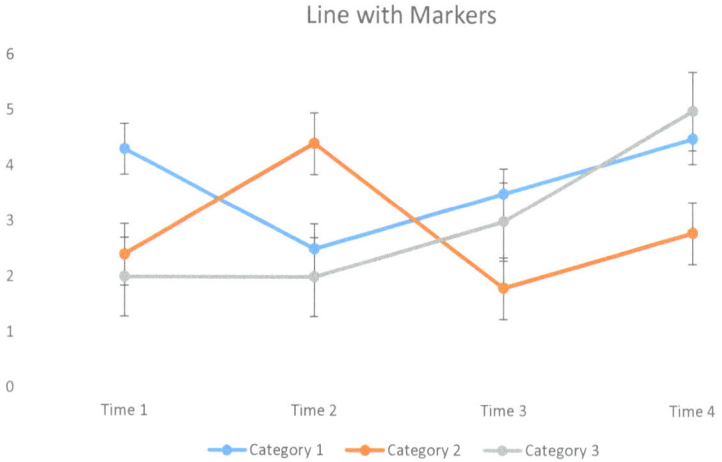

Figure 27: Line chart of the same data as in Figure 26, but with markers and standard error bars for actual data values

I think charts of this form would need to be used with caution, and I would recommend a Stacked Column or Stacked Bar chart be used instead.

Area charts

Another set of charts which could be considered for use in similar situations are the Area Charts, available as simple (unstacked) charts such as the 2D ones in Figures 30 and 32 and their 3D equivalents as in Figures 31 and 33, plus Stacked Area charts such as the 2D one in Figure 34 and its equivalent 3D version in Figure 35.

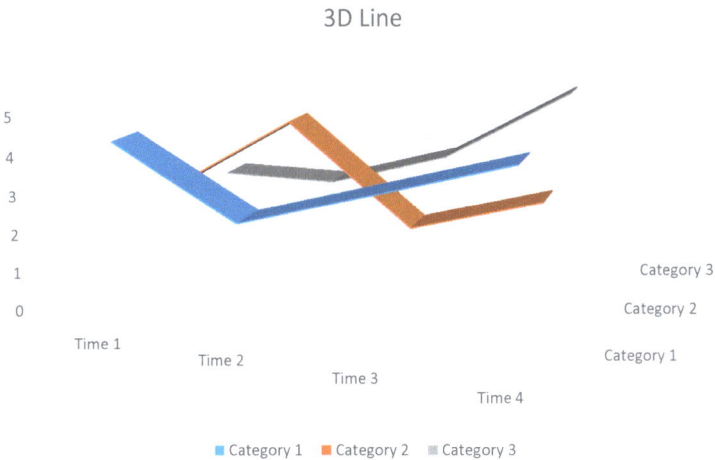

Figure 28: 3D Line Graph, with same data as in Figures 26 and 27

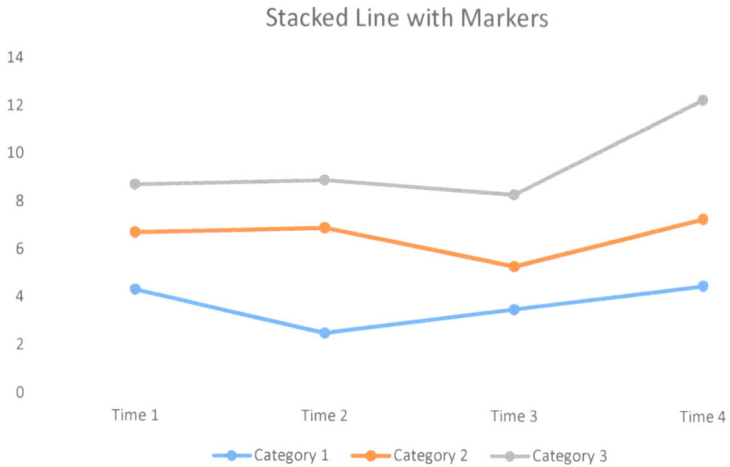

Figure 29: Stacked Line chart of the data in Figure 27, again with markers for actual data points

The main fault displayed in Figure 30 is obvious at first glance – where is the missing data for the last two readings of Category 1, i.e. the blue coloured area? It is masked by the red area of Category 2, which has the higher readings at these last two times. Perhaps if front colours are made transparent, colours behind may be acceptably visible. The same fault exists in Figure 31, except that it is some of the red area that is hidden, rather than the blue. For this reason I would not recommend either version of this style of graph.

The other possible fault in Figure 30 is that, if the blue area is meant to be the most important information, as perhaps indicated by the legend which shows blue as representing Category 1 (as used throughout the other diagrams in this book), why is it not placed in the lowest position close to the horizontal axis, with it in front of Category 2 data? Therefore, if any data points are hidden, they will only be those of less important categories.

This 'fault' has not been repeated in Figure 31, which in fact displays the blue areas of the Category 1 missing in Figure 30, but of course at the expense of hiding some of the red areas.

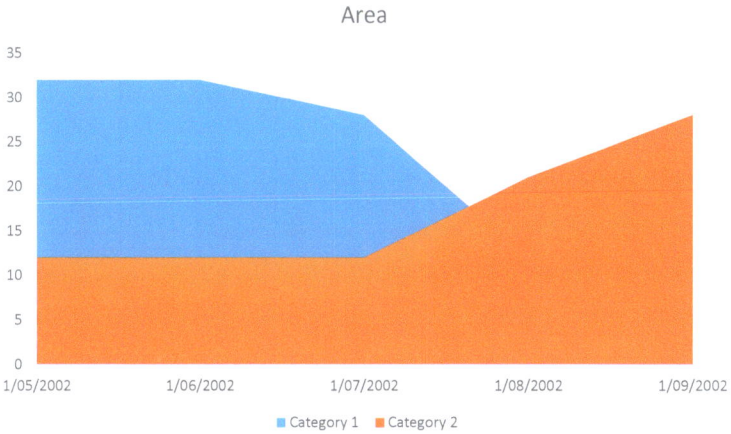

Figure 30: Area graph of two categories over time

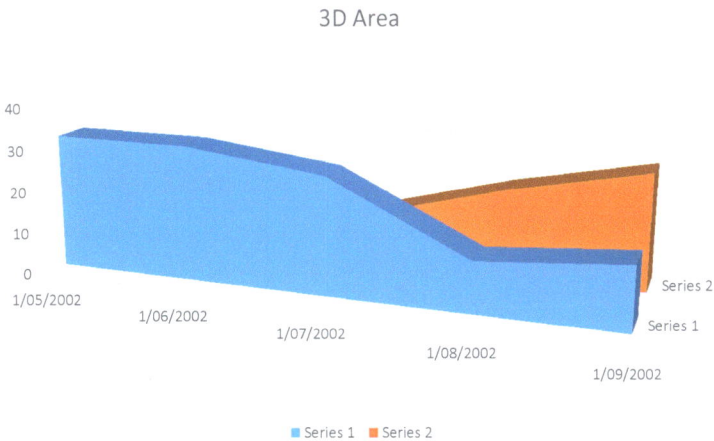

Figure 31: 3D Area chart of two categories over time

Area charts with 'transparent' colours do exist, as seen in Figures 32 and 33, which are alternate versions of Figures 30 and 31 respectively, but they are still a little difficult to read.

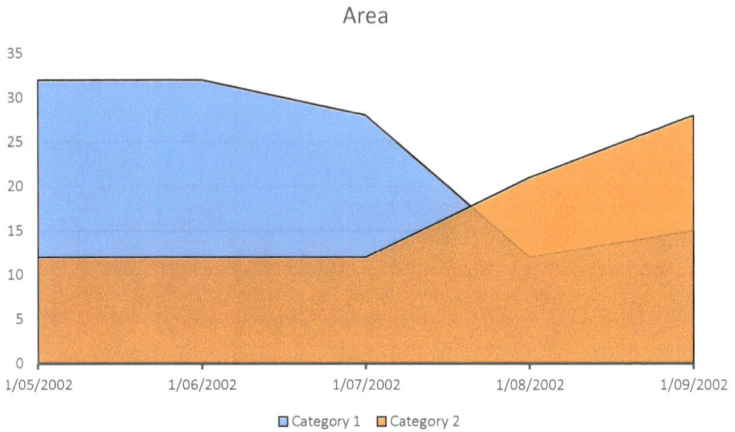

Figure 32: Area graph of two 'transparent' categories over time

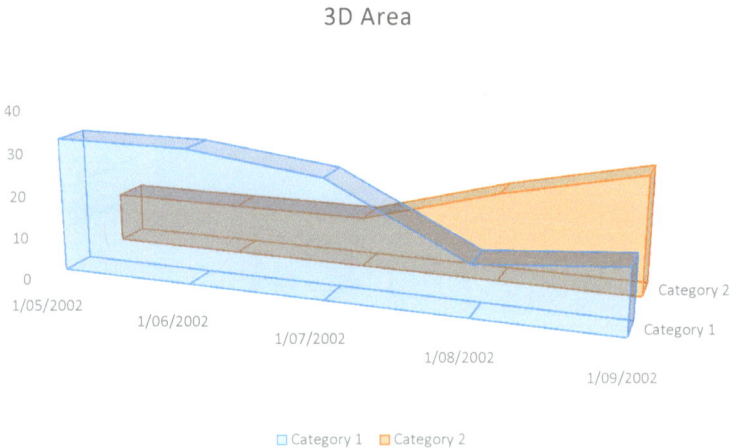

Figure 33: 3D Area chart of two 'transparent' categories over time

Stacked area charts

In Figures 34 and 35, which are stacked area charts, there are no hidden data points of either of the categories. If we assume that the blue area represents the values for the most important category, then the categories have been placed in the preferred order and should match the discussion of results if they too are arranged in the preferred order from most important to the least important.

One fault in these last two Figures is that it is difficult to ascertain the correct grid readings at the junction of the two different coloured areas. This is actually a fault with all of the area graphs in Figures 30-35, as so much of the gridlines is obscured by the areas or volumes. Basically, I don't believe that area charts have much to offer when you see the possible problems which can arise.

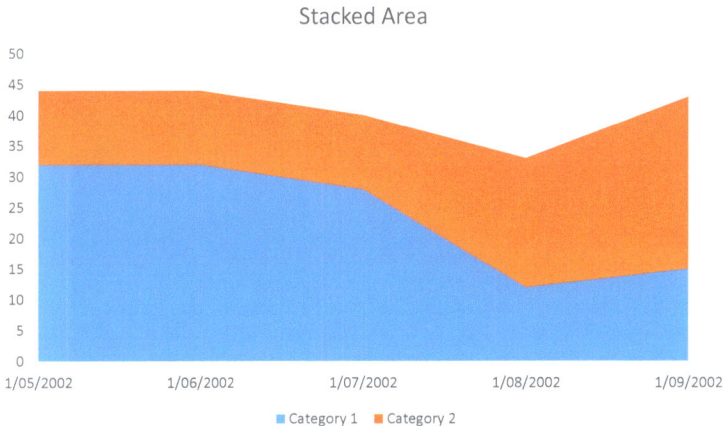

Figure 34: Stacked Area graph of two categories over time

3D Stacked Area

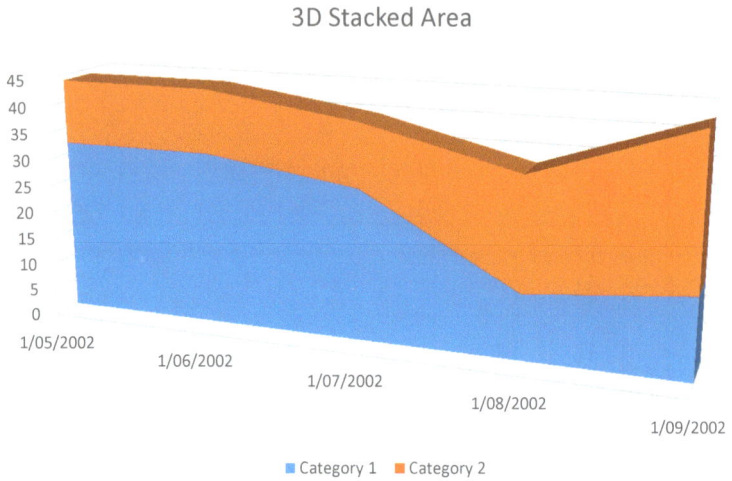

■ Category 1 ■ Category 2

Figure 35: 3D Stacked Area graph of two categories over time

As mentioned much earlier, it is heights, not areas or volumes which are plotted in Figures 30-35, and also as mentioned before, the quadrilateral areas displayed between the data points may give a false impression that changes between data points are linear, whether this is true or not. Column or Bar charts don't give this impression, so they should be used, unless there is some proof that changes between data points are in fact linear.

Again, it is to be hoped that your pictures will be easy for your readers to understand and interpret correctly so they are indeed worth more than any words which would otherwise be required to be written or, more importantly, to be read.

Chapter 5

Using Scatter Plots with and without Lines

Two-dimensional and three-dimensional plots

To see if two variables are related, the data can be plotted as 2-dimensional coordinates on two axes at right angles, say X along the horizontal axis and Y along the vertical axis. In fact, if there is a third related factor, the data can be plotted as 3-dimensional coordinates with a third axis (the Z axis) at right angles to both other axes.

Typically to draw three axes on paper or on screen (i.e. to draw in 3D on a 2D surface), X is again along the horizontal axis, Z is along the vertical axis and Y is along an axis which simulates coming out of the paper by being drawn at an angle such as 135° to each of the other two axes, as illustrated in Figure 36. Each line passes through the origin and continues for the plotting of negative values.

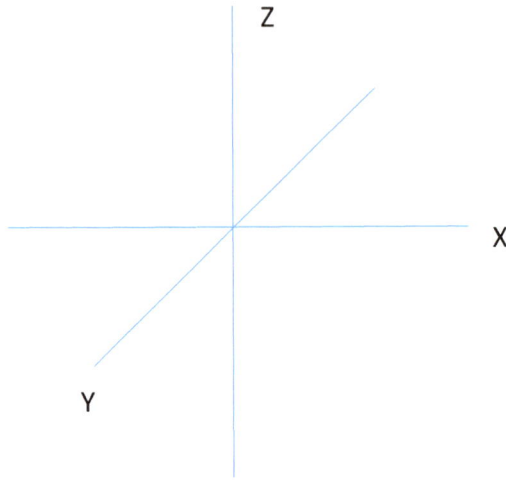

Figure 36: 3-Dimensional coordinate system

3D plotting on three axes will not be further discussed in this book, so the rest of this chapter deals only with plotting on two axes.

Scatter plots for two variables

Figure 37 has been deliberately drawn to demonstrate a situation whereby Y is obviously not related to X or, at best, very weakly related, so no valid trend line, straight or curved, can be drawn through the data. In essence, any value of X can give you a whole range of Y values, and *vice versa*.

This situation is in vast contrast to the opposite situation in Figure 38, where a clear relationship exists, even before any trend line is drawn or calculated with statistical techniques such as regression. By the way, I think it is highly desirable to always plot the actual values as markers, even if an estimated or a calculated line has been drawn, as the markers help to display the spread of results and therefore the errors surrounding the line.

Correlation and regression

Figure 39 uses the same data as in Figure 38, but a trend line has been added (manually) to roughly indicate a possible mathematical relationship between the two variables. If correlation and linear regression analysis had been conducted, then the actual mathematical relationship between the two variables would be contained in the equation for the linear regression line drawn.

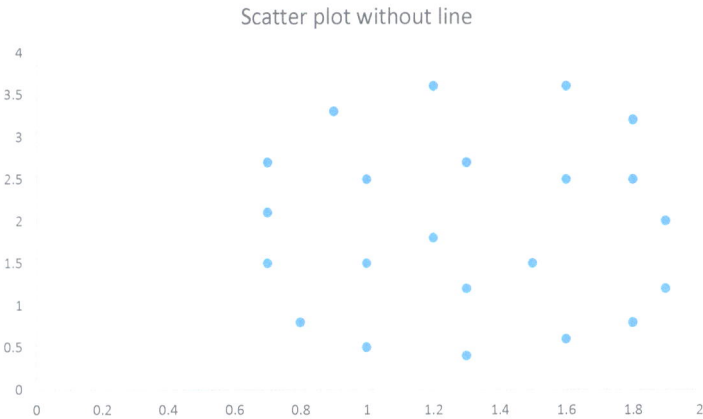

Scatter plot without line

Figure 37: Scatter plot without any valid relationship, so no trend lines

Scatter plot without line

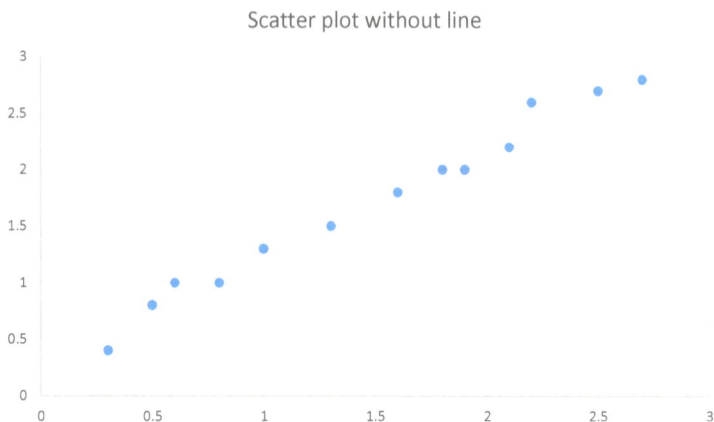

Figure 38: Scatter plot with fairly clear relationship,
even without any trend line drawn

If the trend line through the scatter plot suggests that it is
curved rather than straight, as shown in Figure 40, then different statistical analysis would determine the equation for
the curvilinear regression line (or another more fitting line
as determined by the analysis) to be drawn to represent the
actual mathematical relationship between the variables.

Scatter plot with straight trend line

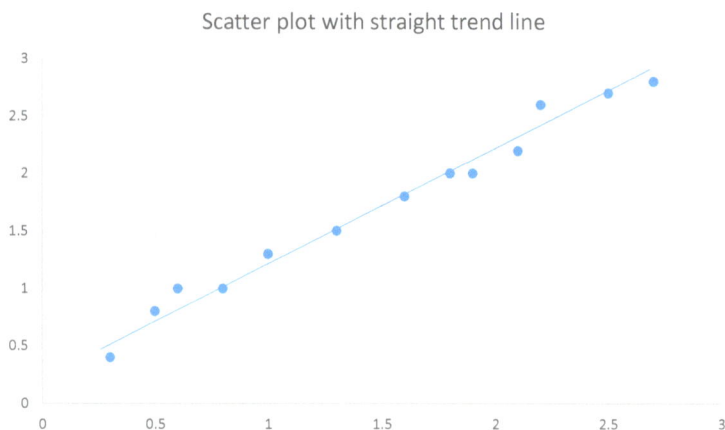

Figure 39: Scatter plot with a linear relationship
between X and Y, and a straight trend line

Scatter plot with curved trend line

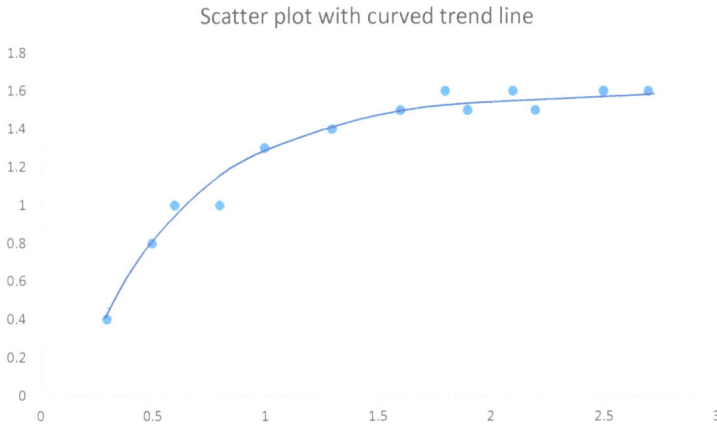

Figure 40: Scatter plot with curvilinear relationship between X and Y, and curved trend line

Bubble charts

I said before that I would not be discussing plotting in 3 dimensions (i.e. with 3 axes), but there are a few ways in which an extra dimension can be portrayed in 2D without introducing another axis. But these can only be effectively used if the data plotted is fairly sparse, as the diameter of the 'dots' will be increased so 'dots' could overlap if they are too close to each other.

These are called Bubble charts, as the third dimension is the diameter or more often the area of the sizes of circles or spheres which replace the normal small dots to represent the plot of the first and second dimensions. Figure 41 shows a series of 'flat' circles of varying size and Figure 42 shows the same series but rendered as 'spheres'. Note that in both of these diagrams, even with the spheres, it is the circular area, not the radius, diameter or volume that indicates the third dimension size, and this point needs to be spelt out in the text accompanying the graph and in the legend of the graph.

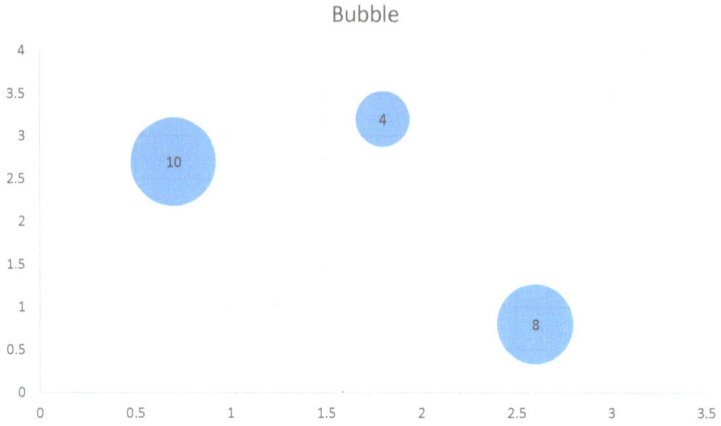

Figure 41: Bubble chart showing third dimension as the size (area) of the plotted circles

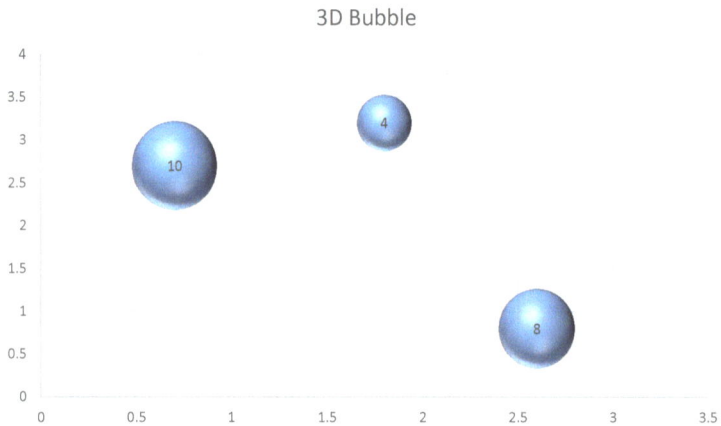

Figure 42: Bubble chart with third dimension as the size (area, not volume) of the plotted spheres

The diagram with the 'spheres' looks attractive, but may exaggerate the differences between the 'spheres' and suggest to the reader that the sizes plotted are linked to the 'volume' of the spheres, whereas they are really linked to the areas on the page or screen. For this reason, I think the 'flat' circles in Figure 41 are a better choice.

Bubble charts are often seen used to indicate human or animal populations of cities or regional areas plotted against latitude and longitude or some other geographical measures. As you can see, if there are too many data points, the circles or spheres will overlap and some will become hidden behind others.

Other alternatives for bubble charts

Another variation of this approach could be to use different colours to indicate different ranges of the size dimension, e.g. red could indicate 0–9, blue could indicate 10–19, green could indicate 20–29 and purple could indicate 30–39, etc. This technique could accommodate more data points as the circle sizes could all be small and all with the same size radius, diameter and area.

This concept is illustrated in the simple coloured bubble chart in Figure 43, where colour, not area, denotes the size in categories.

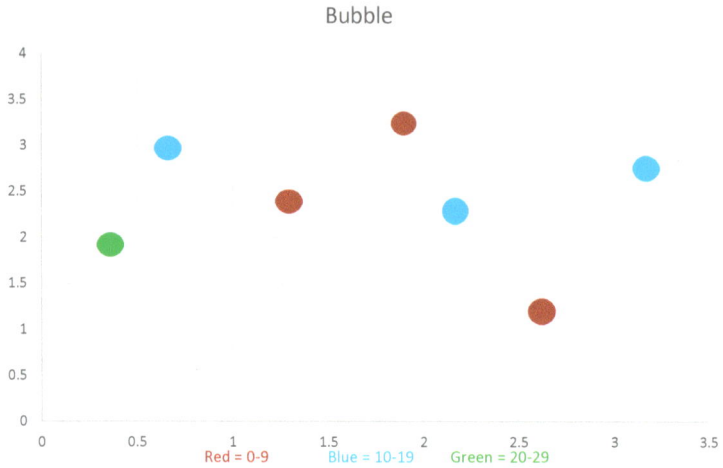

Figure 43: Bubble chart with colours depicting the size ranges of the third dimension categories

Other versions of this approach could be to use different shapes for the categories or to use both colours and shapes combined to represent categories. Figure 44 uses shapes to represent the categories, while Figure 45 uses the combination of colours and shapes to represent categories.

This last version looks far too busy (and a bit of overkill) to me, so if shapes are to be used for categories I would recommend Figure 44 rather than Figure 45.

However, because each of the different shapes in Figures 44 and 45 give the impression that they have different areas even if they don't, using circles of uniform area as in Figure 43 is definitely better than Figures 44 and 45.

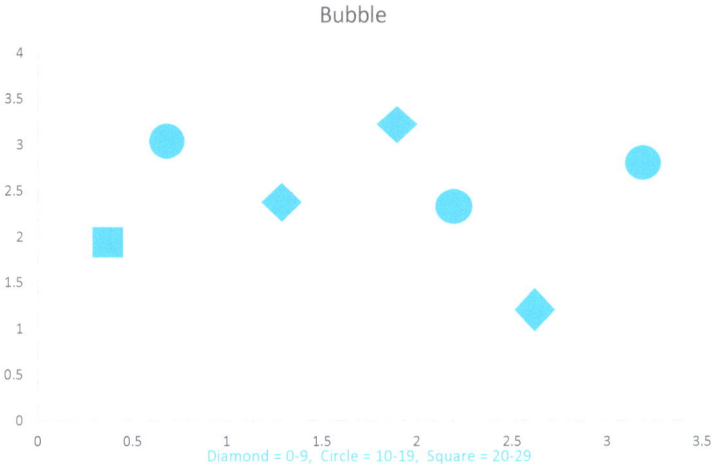

Figure 44: Bubble chart with shapes depicting the size ranges
of the third dimension categories

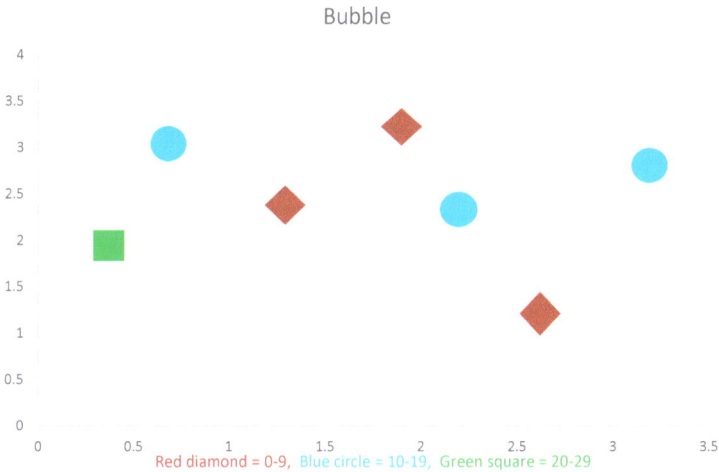

Figure 45: Bubble chart with combinations of colours and shapes
depicting the size ranges of the third dimension categories

Whatever graph you choose, make sure the picture you present sells your message properly and is worth many, many words.

Chapter 6

Plotting in More than Three Dimensions

Multiple dimensions

At times you will be working with multiple dimensions, i.e. more than three dimensions, and need a chart designed for this purpose. A good example of work with multiple dimensions is taste panel or sensory evaluation panel assessment of a food product, where properties such as texture, moisture content, brittleness, crunchiness, visual appeal, colour, colour intensity or shade, odour, sweetness, sourness, salt content, bitterness and other characteristics are judged by a group of people.

The panels may be consumer groups assessing the appeal of new products being developed or they can be trained assessors skilled in discriminating differences in the characteristics to fine tune the development of the products. Because of the variability inherent in people's appreciation of foods, statistical design and analysis will be essential to gain realistic results.

Radar or Spider Web charts

So often there are more than three factors being judged in a trial. The special chart designed for recording such work in multiple dimensions is called a radar chart or spider web chart, as illustrated in Figure 46.

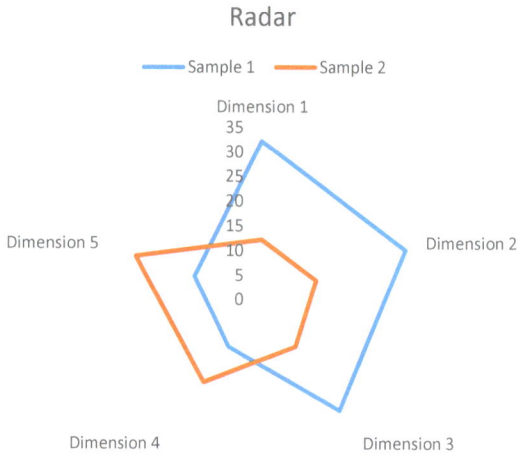

Figure 46: Radar chart of two samples tested against five dimensions

Here two samples have been assessed on five dimensions. The blue lined Sample 1 has scored 32 on Dimensions 1 and 2, 28 on Dimension 3, 12 on Dimension 4 and 15 on Dimension 5. The red line Sample 2 has scored 12, 12, 12, 21 and 28 respectively on the same five dimensions.

Perusal of the graph shows that Sample 1 was better than Sample 2 in the first three dimensions but Sample 2 was better than Sample 1 in the other two dimensions.

I think a better version of Figure 46 is Figure 47, showing the same scoring of both samples over the same five dimensions, but with the data points marked by dots on each of the dimension scales.

The lines connecting the dots have no actual significance other than helping to identify the samples by linking the data points with the same colour. Also, if the lines in Figure 46 were eliminated there would be no data in the graph at all.

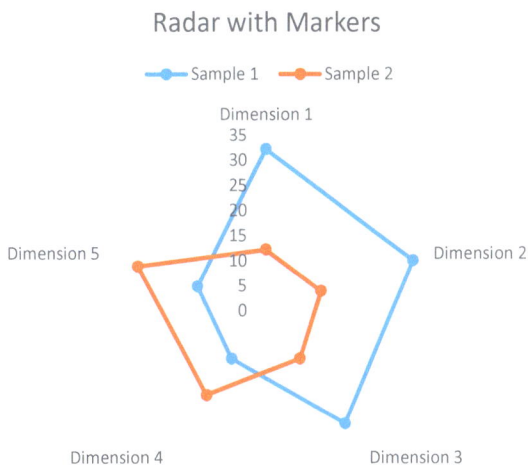

Figure 47: Radar chart of two samples tested against five dimensions, with data points marked

Extra dimensions and sample data

While these two diagrams show the sample data plotted on only five dimensions, radar charts can be used for many more dimensions. Figure 48 shows the same two samples with extra data plotted against another two dimensions.

As you can see, increasing the number of dimensions scored adds more complexity to the diagram, but increasing the number of samples adds even more clutter, as demonstrated in Figure 49, with an extra Sample 3.

Too much data can be confusing for the reader. Once again, display your most important information first and carefully, so if you have six samples they may be best illustrated in two or three diagrams. Don't spoil your picture or you might need hundreds of words to describe it.

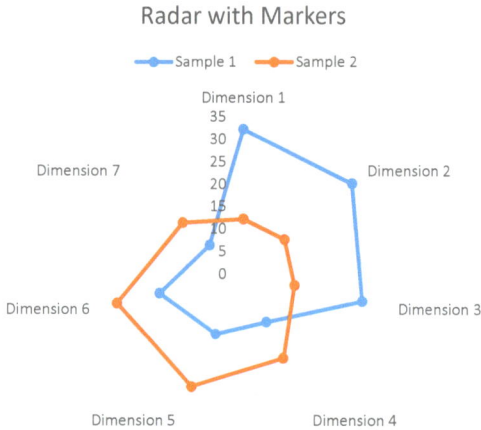

Figure 48: Radar chart of two samples tested against seven dimensions, with data points marked

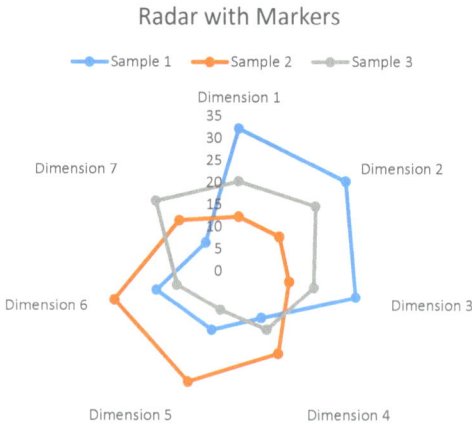

Figure 49: Radar chart of three samples tested against seven dimensions, with data points marked

Filled radar charts

There is another radar chart type where sample areas are filled with colours, as in Figure 50, which uses the same data from Figures 46 and 47. This suffers from the same problem discussed with area charts in earlier chapters – some of the data points are hidden. Not only that, but the coloured areas also obscure the grid lines in the background.

The filled areas serve no purpose other than to link the data points of each sample under the same colour, so they have no more value than the linking lines in the earlier radar charts. I see no real value in these filled charts and would not recommend using them.

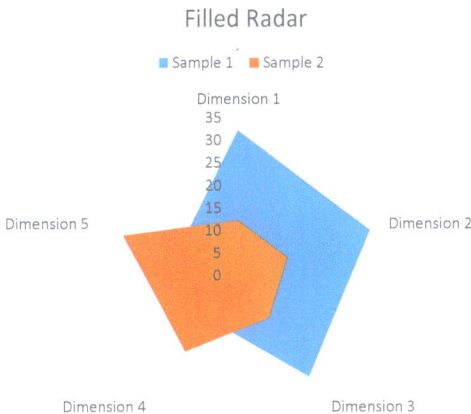

Figure 50: Filled Radar chart of two samples tested against five dimensions

Also, assuming the colour scheme adopted in this book has blue as the colour of the most important sample and confirmed with blue for the first sample in the legend, the data hidden here is part of the most important sample.

However, there is an alternate version of Figure 50 which uses 'transparent' colours so all data points are visible. This is Figure 51, which has the same data as Figures 46, 47 and 50.

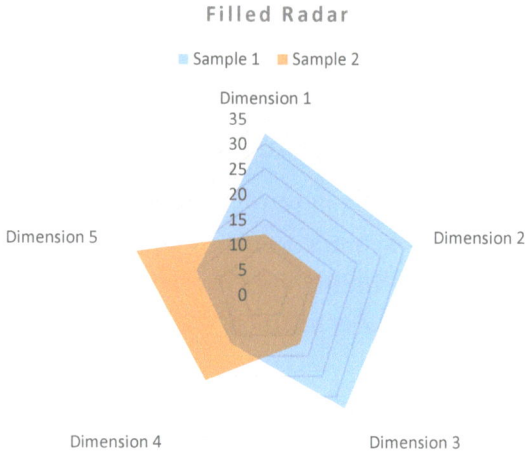

Figure 51: Transparent Filled Radar chart of two samples tested against five dimensions

Chapter 7

Plotting Special Data Including Stock Market Data

Charts for stock market information

There are special charts designed for reporting stock market information.

The simplest example is a plot of High values, Low values and Closing values, as displayed in Figure 52, which does not have the Opening values or the Transaction Volumes. The tops of the thin vertical bars are obviously the High values and the bottoms of the bars are the Low values. The grey dot markers on the bars are the Closing values.

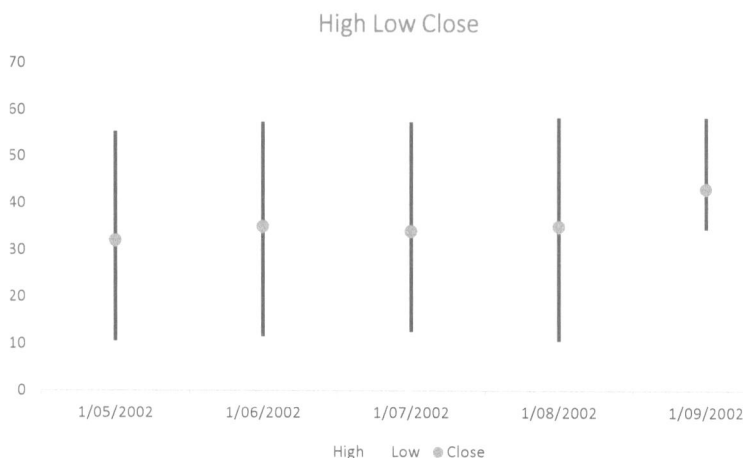

Figure 52: A plot of High, Low and Close Values for
a Stock Item over Different Periods

The next example includes a plot of Opening values as well as those above, as shown in Figure 53, but it still does not display the Transaction Volumes. Again, the tops and bottoms of the thin vertical bars are the High values and the Low values respectively.

The boxes on the vertical bars show the Opening and Closing values. If the box is white, then the bottom of the box is the Opening value, which rises through the values in the box to the Closing value at the top of the box; if the box is black it means the Opening value is at the top of the box which falls through the values in the box to the Closing value at the bottom of the box.

Although I am not very familiar with stock market financial data, it seems to me that the addition of Opening values would be both desirable and easy to do, so I would recommend the format in Figure 53 over that in Figure 52.

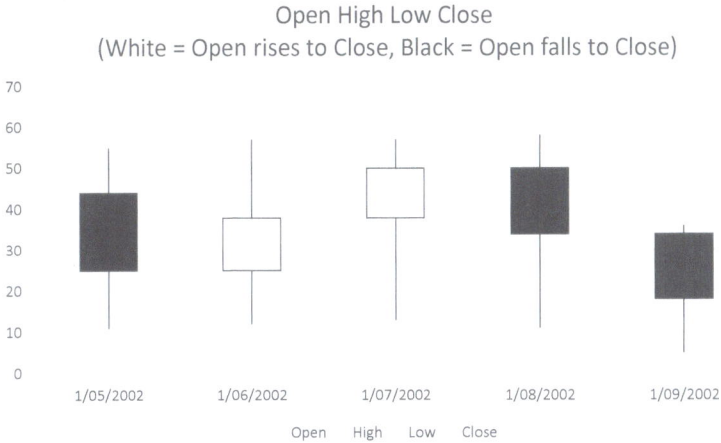

Open High Low Close
(White = Open rises to Close, Black = Open falls to Close)

Figure 53: A plot of Open, High, Low and Close Values for
a Stock Item over Different Periods

An extra component, Transaction Volumes, can be added to
Figure 52 or to Figure 53 to create the diagrams displayed in
Figure 54 or Figure 55 respectively.

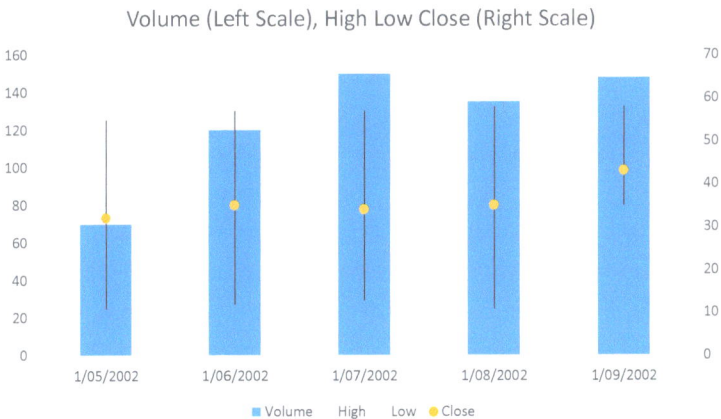

Volume (Left Scale), High Low Close (Right Scale)

Figure 54: A plot of Volume, High, Low and Close Values for
a Stock Item over Different Periods

Again, Figure 54 is the simpler of the two as it does not have the Opening values. Again, the tops and bottoms of the thin vertical bars are the High and Low values and the yellow dot markers on the vertical bars are the Closing values. The transaction Volumes are the blue columns, the values of which are to be read off the scale on the left, while all three of the other factors, Highs, Lows and Closing values are to be read off the scale on the right.

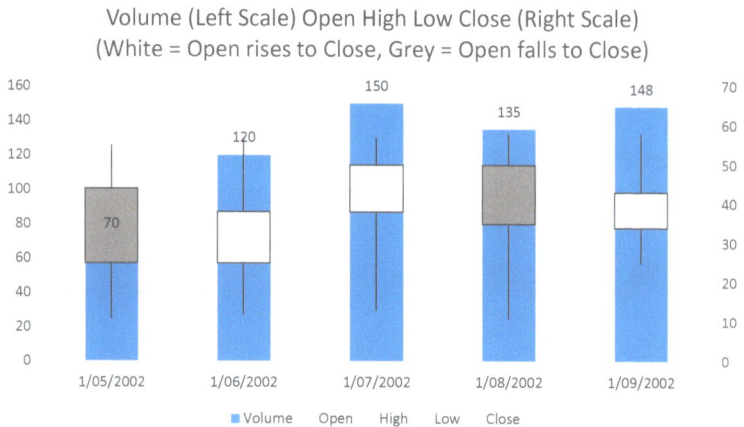

Figure 55: A plot of Volume, Open, High, Low and Close Values for Stock over Different Periods

Figure 55 has all five value types, Transaction Volumes, Opening values, High values, Low values and Closing values. Again Volumes are the blue columns with their scale on the left, while the other factors use the scale on the right. Again the tops and bottoms of the thin vertical lines are the High and Low values respectively.

Again, the Opening and Closing values are shown by the tops and bottoms of the white and grey boxes. If the box is white, then the Opening values are at the bottom and they indicate an overall rise to the Closing values at the top; if the box is grey, then Opening values are at the top and fall to the Closing values at the bottom.

Because the top of the first blue column is in an intermediate position within the corresponding Open-Close box, I have included black figures for the Volume readings, which appear above each blue column. However because using black as the box colour for falling values (as in Figure 53) would obscure the first Volume reading of 70, I have chosen mid-grey for the colour of the falling values box.

Because Figure 55 has extra useful information in it, I would prefer it over Figure 54, even though in this particular case, the typical black colour traditionally used for falling values boxes had to be replaced by mid-grey to display the black Transaction Volume values. Another approach would be to leave the falling Open-Close boxes black and use any colour other than black for the Volume values, as has been done with the green values in Figure 56.

Of all the stock market information graphs presented here, I would favour using Figure 55 or Figure 56, as they contain the most information. Because of my limited knowledge of stock market transactions, I cannot realistically comment on which of the five values would be the most important, but I would assume that all of the values are important and Figures 55 and 56 display them all.

Box and whisker charts

There is another graph type called a Box and Whisker Chart, as shown in Figure 57. The Whiskers can have different uses depending on the needs, but here they mark the maximum and minimum values.

Volume (Left Scale) Open High Low Close (Right Scale)
(White = Open rises to Close, Black = Open falls to Close)

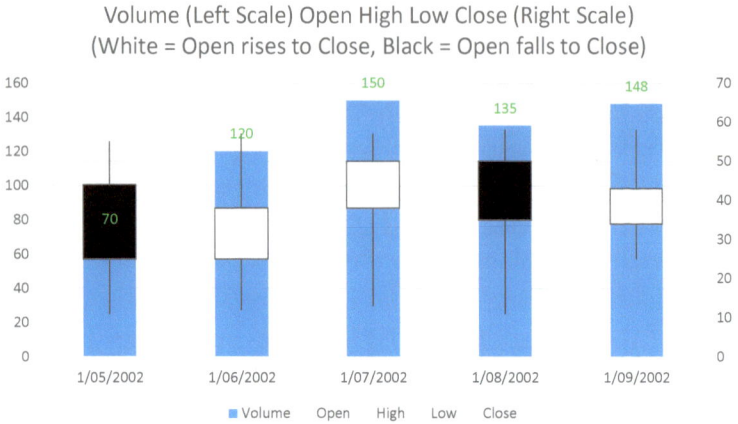

Figure 56: Alternative plot of Volume, Open, High, Low and
Close Values for Stock over Different Periods

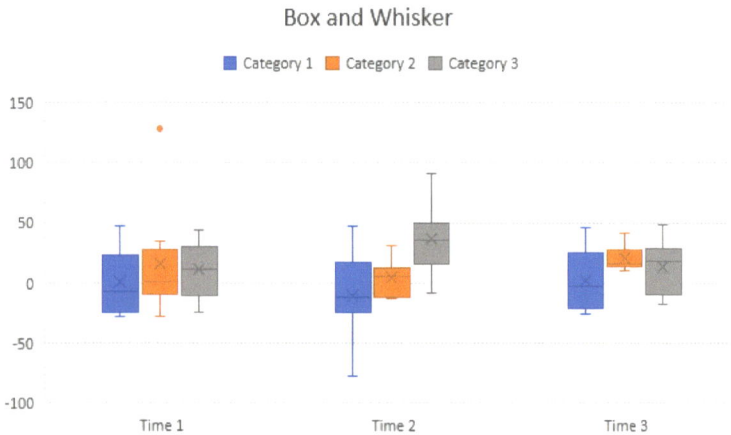

Box and Whisker

Figure 57: Box and Whisker Chart Showing the Highest, Lowest,
1st, 2nd and 3rd Quartile Boundaries, Median and Mean Values

The values plotted in Figure 57 are the Highest, the Lowest,
the 1st Quartile upper boundary, the 2nd Quartile upper

boundary which is also the Median (middle value), the 3rd Quartile upper boundary and the Mean.

The 2nd and 3rd Quartiles are contained in the boxes, with their common boundary marked with a horizontal line within the box. The Mean is marked with a cross inside the box. The Highest and Lowest values are at the tick marks at the ends of the thin vertical lines extending up and down from the boxes.

The lowest 25% of the values form the 1st Quartile. The next 25% of the values form the 2nd Quartile. The next 25% form the 3rd Quartile, while the last 25% is the 4th Quartile. The Median is the middle value of all values and marks the boundary between the 2nd and 3rd Quartiles; this is different from the Mean (actually the Arithmetic Mean) which is the arithmetic average of all the values.

The dot above the red Category 2 at Time 1 represents one value which is well away from all the rest of the Category 2 Time 1 values, so although strictly it is the maximum value obtained, it is considered an outlier or an abnormal reading and may not be considered a valid result to be discussed.

This style of graph contains a lot of information and attempts to provide a picture of the spread of the results and how well the Median or the Mean could represent the range of values. Naturally, half of the values at a given Category and Time will lie within the boxed portion. Generally speaking, here the red Category 2 values are not as spread out as those of the blue Category 1 or grey Category 3 values.

Waterfall charts

The next graph to be considered is the so-called Waterfall Chart as displayed in Figure 58. This indicates the increases and decreases of the values of some variable, often financial information, occurring, typically over time, and the effects these have on the balance, residual value or total value of the variable.

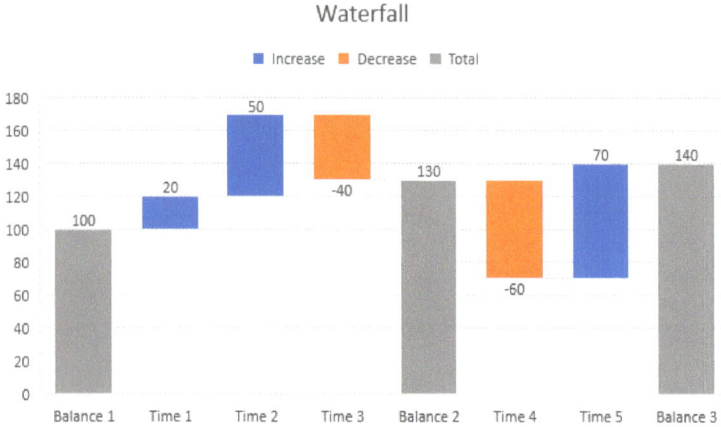

Figure 58: Waterfall Chart Showing Balances, Increases and Decreases over Time

The plot shows the opening balance of 100 is increased twice, by 20 and 50, before being decreased by 40 to bring the middle balance to 130, etc. In monetary terms the increases would be revenue items and the decreases would be payments, so the balances would be the bank balances or book balances at set times, e.g. at each month.

The data does not have to be monetary values, although it often is; the data could instead be inventory stock or hired employees, etc.

Combination graphs

Sometimes a Combination Graph, often colloquially called a Combo Graph, is used to display a range of variables in different formats in the same graph or chart. The most common combination is a mixture of Column Charts and Line Charts in the one Combination Chart, as displayed in Figure 59.

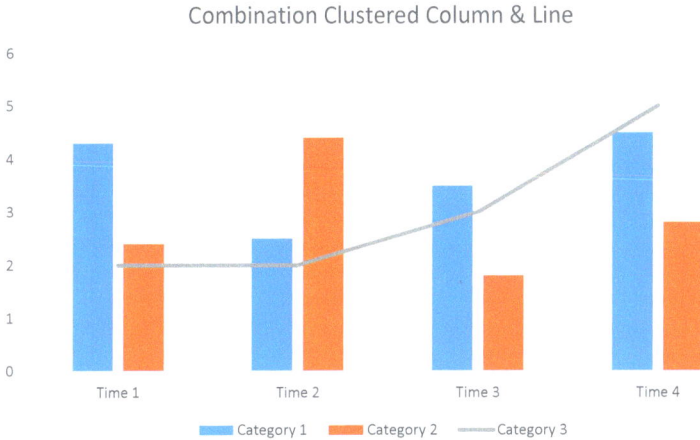

Figure 59: Combination Chart with Two Variables Plotted as Columns and One Plotted as a Line

Most often, if one variable is plotted in a different format from the other variables, it will be a quite different variable from the others. For example in Figure 59, the blue and red Category 1 and 2 columns could be sales volumes of two products over four times, while the grey Category 3 line could be numbers of salesmen over the same four times.

Often the different variable will have its own secondary axis on the right of the diagram, as its scale may be markedly different from the scales used for the other variables.

There are literally thousands of different ways that information can be presented and many of these are especially designed for use in highly specific work areas, business or commercial situations and home, club and entertainment options.

Whatever data is being presented, give considerable thought to how it can be made easier for your readers or audience to read, understand and interpret the information accurately. Remember that data can be facts and figures, but it is only when data is arranged and presented correctly, that it becomes information with valuable messages and can be used.

Chapter 8

Final Comments and Summary

The following dot points summarise all the main aspects of data and information presentation dealt with in this book:

Treatment and Results

- Randomise all the treatments to be dealt with in experimentation or observance of facts.

- Temporarily label all the treatments and other factors to be dealt with before the work starts.

- Once the results are known, re-allocate the order and labels of the treatments and other factors to be able to present them in an optimal order, typically in descending order from most important to least important.

- Write up the experimental methods/techniques/observations in descending order from most important to least important.

- Write up the results in descending order from most important to least important.

- Discuss the results in descending order of importance from most important to least important, so methods, results, tables, charts, discussion and conclusions are all aligned and easy for the reader/audience to link the parts together quickly.
- If there is a choice between positive effects of the treatments, e.g. which one had the lowest number of survivors perhaps or highest growth rate perhaps, and negative effects of the treatments, e.g. which one had the highest number of items removed perhaps or the lowest growth rate perhaps, choose whichever is the realistic approach concerned, e.g. are you interested in increasing growth rate or deceasing it?
- Optimise readability of tables and charts.

Tables

- If tables are to be used, place results in them in descending order of importance, with the most important results in column 1 (and possibly in row 1, if applicable) and the least important in the last column (and possibly in the last row, if applicable).

Pie charts

- If pie charts are involved, always place category results in descending order from most important to least important, starting at the 12 o'clock position for the most important and proceeding clockwise to the least important.
- Preferably place numeric or percentage values in the chart.
- Don't alternate large and small segments around the pie, don't start from other than the 12 o'clock position and don't proceed anti-clockwise around the pie.

- Preferably use colours in pie charts and choose contrasting adjoining colours to distinguish between segments easily.

- If there are multiple coloured diagrams in a document, whether pie charts or other charts, use the same colour scheme throughout the document to aid interpretation.

- If colours can't be used, choose either decreasing shades of grey or black and white striped, dotted or hashing to distinguish between segments easily.

Histograms

- Be careful to form unambiguous boundaries between categories, e.g. (0–5 and >5–10) or (0–<5 and 5–<10), not (0–5 and 5–10) or (<5 and >5).

Stacked Column, Bar, Line and Area charts for relative or percentage values

- Percentages or relative values are plotted as heights, not as areas or volumes, in 100% stacked columns, bars, line or area charts, whether in 2D or 3D versions.

- Place the most important information in the most prominent position – closest to the horizontal axis for 100% stacked columns, lines and areas or closest to the vertical axis for 100% stacked horizontal bars.

- Place the least important information in the least prominent position – furthest from the horizontal axis for 100% stacked columns, lines and areas or furthest from the vertical axis for 100% stacked horizontal bars.

- Place intermediate information in decreasing order between the most important information and the least important information in 100% stacked columns, bars, lines and area charts.

Line charts for relative or percentage values

- Use data point markers in all line charts.
- Be careful using or reading line charts to distinguish between unstacked plots and stacked plots, whether 100% or not, because they can look similar and be interpreted inaccurately.
- Be careful about reading straight lines between data points as indicating linear changes between data points, as this may be invalid.

Area charts for relative or percentage values

- 100% stacked area charts typically have some grid lines hidden behind some data points, so these charts should be avoided or only used with caution.
- Consider using transparent colours to reveal values which would otherwise be hidden.
- Be careful about reading quadrilateral areas between data points as indicating linear changes between data points, as this may be invalid.

Column, bar, line and area charts for actual values

- Actual values are plotted as heights, not as areas or volumes, in column, line and area charts or as lengths in horizontal bar charts.
- Standard error bars can be plotted up and down from the actual values in column, line and area charts or left and right from the actual values in horizontal bar charts to display spread of the data.
- In stacked actual values charts, place the most impor- tant information in the most prominent position – closest to the horizontal axis for stacked column, line

and area charts or closest to the vertical axis for stacked bar charts.

- In stacked actual value charts, place the least important information in the least prominent position – furthest from the horizontal axis for stacked columns, lines and areas or furthest from the vertical axis for stacked horizontal bars.

- Place intermediate information in decreasing order between the most important information and the least important information in stacked columns, bars, lines and area charts.

- In stacked actual values, the tops of the stacked columns, lines and areas or the right hand ends of stacked bars indicate the total values at each data point.

Column charts and bar charts for actual values

- In some 3D versions, some data points may hide behind other values, so 3D should be used with caution.

- In some 3D versions, it may be difficult to read some values accurately, because of distance from the background grid.

Line charts for actual values

- Use data point markers in line charts
- Be careful using or reading line charts to distinguish between unstacked plots and stacked plots, whether 100% or not, because they can look similar and be interpreted inaccurately.
- Be careful about reading straight lines between data points as indicating linear changes between data points, as this may be invalid.

Area charts for actual values

- Some area charts typically have some data points hidden behind other data points, so should be avoided or only used with caution.

- Consider using transparent colours to reveal values which would otherwise be hidden.

- Be careful about reading quadrilateral areas between data points as indicating linear changes between data points, as this may be invalid.

Scatter or X–Y plot diagrams

- Scatter or X–Y plot diagrams can reveal whether there is a trend line for the data or not.

- If a trend exists, it may be a straight line or a curved line, indicating a linear or a curvilinear relationship between X and Y.

- Correlation and regression analysis will indicate whether any relationship is valid and will assist in plotting the line which best fits the relationship.

Bubble charts

- A third dimension can be indicated in a 2D plot by using filled circles of varying area as data points, where the circle areas represent the value of the extra dimension.

- Pseudo-spheres of varying area when projected on the 2D surface can be used instead of circles, but it is still the area, not the volume that represents the extra dimension value.

- Third dimension values can also be indicated by plotting data values or ranges as differently coloured dots, differently shaped 'dots' or by a combination of different colours and shapes.

Radar or Spider web charts

- More than three dimensions can be plotted in radar charts or spider web charts.
- Preferably use markers for the data point values along each axis.
- These charts can become cluttered if more than three sets of data are plotted.
- Filled versions of these exist but unless the colours are transparent, some data points can become hidden.

Stock market information charts

- The simplest stock market and financial information charts include highest values, lowest values and closing values.
- More complete stock market charts also include opening values or sales volumes or both extra values.
- Rises and falls between opening and closing values are indicated by different coloured boxes, typically white for rising values and black or some other dark colour for falling values.
- When volume is represented as filled coloured columns, some other values, especially the boundaries of the open-close boxes, may become hidden, unless either a light grey is used to fill the falling open-close boxes or the value figures are printed in a contrasting colour.

Box and Whiskers charts

- Box and whiskers charts are designed to display spread of results without strong statistical analysis, typically showing highest value, lowest value, 1st quartile upper boundary, 2nd quartile upper boundary, 3rd quartile upper boundary, median and mean; the last five values are contained in a box, while the other two are above or below the box.

Waterfall charts

- Waterfall charts display balances, typically at set times, together with successive increases and decreases between the set times, which typically gives a picture of cash flow, inventory flow or some other feature flow over time.

I hope this book has been of assistance to you in considering how to present your information in the most effective way. There are of course many techniques which have not been described or discussed in this book, so there are bound to be many choices that will suit your information. I wish you well in constructing your pictures to suit your information and your readers or audience.

Bibliography

The concepts outlined in this book are mainly the work of the author, gained from over 40 years of working as a scientist and manager in food science and technology, including writing, statistical analysis, editing and presenting papers and major reports.

For examples of the various presentation formats, I have drawn heavily on the charts and other presentation vehicles publicly available from the following sources, but I have modified much of data in them to specifically illustrate each of the points I make:

Anonymous, (2013–2018), Several on-line articles on aspects of statistics, tables, graphs, charts, in *Wikipedia®*: *The Free Encyclopedia*, Wikipedia Foundation Inc Encyclopedia On-Line, US.

Anonymous, (2013–2016), Several of the charts and graphs in Microsoft® Word 2013, Microsoft® Word 2016, Microsoft® Excel 2013 and Microsoft® Excel 2016 Programs, within

Microsoft® Office 2013 and Microsoft® Office 2016 Software Packages, Microsoft Corporation, US.

Wang, Wallace, (2013), *Microsoft® Office 2013 for Dummies*, John Wiley & Sons, Hoboken, US.

INDEX

About the Author

With a career spanning over 40 years in Queensland State Government Food Science and Technology laboratories in Australia (including eight years in a regional food laboratory in Toowoomba and the remaining period in the main food laboratories in Brisbane), as well as a Master's Degree in Science (Microbiology), a Postgraduate Diploma in Information Processing and a Postgraduate Diploma in Business Administration, Bill Dommett is highly qualified and very experienced in science, computer science and management.

During his career, Bill's work and responsibilities ranged from middle management, team leader, scientific research, quality compliance analysis and assurance, tertiary institute lecturing and industry training, supervision of tertiary students' Master's degrees, and computer systems analysis and design.

Throughout his career, Bill was heavily involved in preparing, writing, statistical analysis, editing and presentation of scientific information from his own and others' work. These included research papers, advisory articles for industry, lectures to tertiary students and industry personnel, seminar and conference papers, reports, standard methods development and work manuals. In addition, he has significant experience in quality control, quality assurance and systems analysis and design of scientific industry programs with and without computers.

While editing scientific research papers, Bill became aware of many situations where improvements could be made to the associated tables, graphs and diagrams to make them more effective tools for communicating the key aspects of the work to the readers or audience. He was also aware of the need to match the text in the methods used, as well as the discussion of results and the conclusions to be drawn with the text and diagrams in the presentation of the results in a logical, easy to follow style.

Bill has now retired and lives in Brisbane, Australia. He has two adult children and four adult grandchildren in Brisbane and in Ohio, US. His current interests include woodcraft, social clubs, travel, fiction and non-fiction reading, documentaries and film.

www.ingramcontent.com/pod-product-compliance
Lightning Source LLC
Chambersburg PA
CBHW041311210326
41599CB00003B/59